CHEM2

NCEA LEVEL 2

DEBORAH HAY

Chem2 NCEA Level 2
1st Edition
Deborah Hay

Cover designer: Cheryl Rowe, Macarn Design
Text designer: Cheryl Rowe, Macarn Design
Production controller: Siew Han Ong
Reprint: Magda Koralewska

For product information and technology assistance,
in Australia call **1300 790 853**;
in New Zealand call **0800 449 725**

For permission to use material from this text or product, please email **aust.permissions@cengage.com**

National Library of New Zealand Cataloguing-in-Publication Data

Hay, Deborah.
Chem 2 / Deborah Hay.
Includes bibliographical references.
ISBN 978-0-17-026010-7
1. Chemistry—Examinations—Study guides. 2.Chemistry—Problems, exercises, etc. 3. Chemistry—Textbooks. I. Title.
540.76—dc 23

Cengage Learning Australia
Level 7, 80 Dorcas Street
South Melbourne, Victoria Australia 3205

Cengage Learning New Zealand
Unit 4B Rosedale Office Park
331 Rosedale Road, Albany, North Shore 0632, NZ

For learning solutions, visit **cengage.co.nz**

Printed in Australia by Ligare Pty Limited.
3 4 5 6 7 8 9 23 22 21 20 19

contents

CHAPTER ONE

2.4 Demonstrate understanding of bonding, structure, properties and energy changes (91164)

Learning outcomes

Tick off when you have studied these ideas in class and when you have revised that section prior to your assessment.

	Learning Outcomes	In class	Revision
1	Explain how ions bond.		
2	Explain how metals bond.		
3	Explain how non-metals bond.		
4	Explain what an intermolecular bond is.		
5	Draw Lewis structures of simple molecules.		
6	Predict the shape and bond angles of simple molecules.		
7	Predict the properties of ionic compounds.		
8	Predict the properties of metallic substances.		
9	Predict the properties of molecular substances.		
10	Predict the properties of covalent network substances – limited to graphite, diamond and silicon dioxide.		
11	State the bonds present and particles that make up a substance.		
12	Explain what an exothermic reaction is.		
13	Explain what an endothermic reaction is.		
14	Given a graph, enthalpy value or experimental observations predict whether a reaction is an endothermic or exothermic reaction.		
15	Calculate the amount of energy released or absorbed from either a mass or number of moles and a change of enthalpy.		
16	State the bonds broken and formed in a given reaction.		
17	Calculate the change in enthalpy of a reaction given bond energy data.		

ISBN: 9780170260107

Pre-test – What do you know?

Section 1: Periodic table

1 What factor is common to all of the elements in a given group of the periodic table?

2 What are the common names of groups 1 and 18?

3 What is a row in the periodic table called?

4 What is a column in the periodic table called?

5 What charge is formed by ions of the elements in groups 2 and 17?

Section 2: Atomic structure

1 Draw the atomic structure of lithium (atomic number = 3; mass number = 7).

2 State how many protons, electrons and neutrons are in a lithium atom.

3 Describe the electron arrangement of lithium.

Section 3: Ions

1 Name the following ionic formulae:

a $CaCO_3$

b $Al(OH)_3$

c KNO_3

2 Write the formulae of the following ions:

a Calcium nitrate

b Aluminium sulfate

c Sodium hydroxide

ISBN: 9780170260107

Atomic structure

The model of the atom was not a simple idea that one person came up with in an hour. It took many years and many different people to work out the structure we know today.

Democritus

The first person in the story was ...

- **460BC Democritus**. He developed the idea of atoms after pounding different substances in a mortar and pestle until only small rounded particles were left. He called these ATOMA – which means indivisible.

John Dalton

Next came ...

- **1808 John Dalton**. He suggested that all matter was made up of tiny spheres that were able to bounce around with perfect elasticity which he called **atoms**. (In other words he rediscovered the same idea.)

Then ...

- **1898 Joseph 'J. J.' Thomson**. He found that atoms could sometimes eject far smaller negative particles called **electrons**.

Joseph Thomson

And then in ...

- **1904**. J. J. Thomson developed the idea that an atom was made up of electrons scattered unevenly within an elastic sphere surrounded by a soup of positive charge, to balance the electron's charge like plums surrounded by a pudding. Known as the 'Plum Pudding Model'.

Ernest Rutherford

Then the important part to New Zealanders ...

- **1910 Ernest Rutherford**. He and his team fired helium nuclei at a sheet of gold foil which was only a few atoms thick. They found that although most of the nuclei went through, about one in 10,000 hit something. Hence the model of the atom was not like a pudding as it had too much empty space. So the idea of a nucleus was born.

Next ...

- **1913 Niels Bohr**. Bohr refined Rutherford's idea by calculating that the electrons were in orbits, rather like planets orbiting the sun. Each orbit contains a set number of electrons.

Niels Bohr

James Chadwick

Finally ...

- **1932 James Chadwick**. To help explain the extra mass that couldn't have been the protons or electrons, Chadwick came up with the idea of neutrons.

So we now know atoms are made of three sub-particles called **electrons**, **protons** and **neutrons**. The protons and neutrons are found inside the central **nucleus** tightly packed and the electrons orbit around the nucleus in defined energy levels or layers.

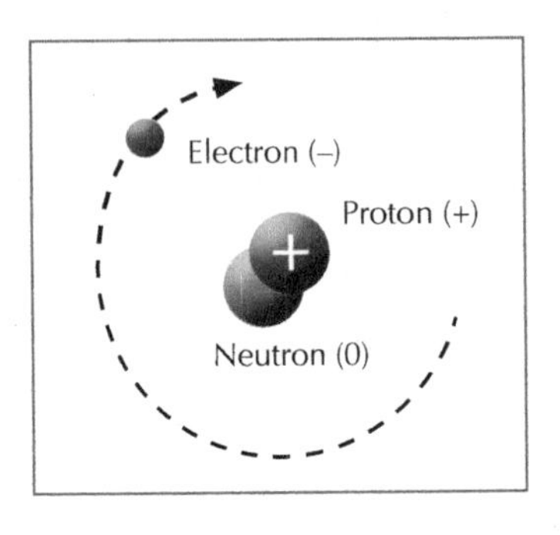

ISBN: 9780170260107

Two important numbers you should remember are:

- The **atomic number** which tells us how many protons an atom has.
- The **mass number** which tells us how many protons and neutrons an atom has (since most of the mass of an atom comes from the protons and neutrons).

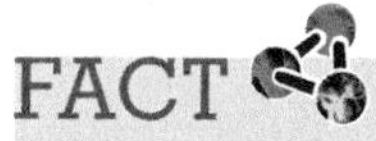

FACT

There are more atoms in the ink in the full stop at the end of this sentence than there are people on this Earth.

The box below you will see frequently throughout this book and it is to help you remember all of the words in chemistry that students commonly confuse.

What do the following key words mean?

Atom	
Matter	
Element	
Compound	
Mixture	
Molecule	

1.1: Atomic structure

Draw the atomic structure of the following atoms using the periodic table on page 183. (The first one has been done for you.)

1 Boron, B	**2** Fluorine, F
3 Aluminium, Al	**4** Neon, Ne

ISBN: 9780170260107

Electron arrangements

- This model shows how the electrons are arranged in an atom.
- In a **neutral** atom there are always the same number of electrons as protons, so the atomic number will tell you how many there are.
- There are always two electrons in the first layer or energy level of an atom.
- There are up to eight electrons in the next layer.
- And finally, there are up to eight* electrons in the last layer.
- When an atom becomes an ion it loses or gains electrons so the electron configuration will change.
- *Example:* when sodium (Na) becomes an ion, its electron configuration changes from 2, 8, 1, to 2, 8 when it becomes Na^+ as it loses an electron.

* *Next year you will find out more about this, but the third layer can have up to 18!*

Atomic structure diagram of carbon

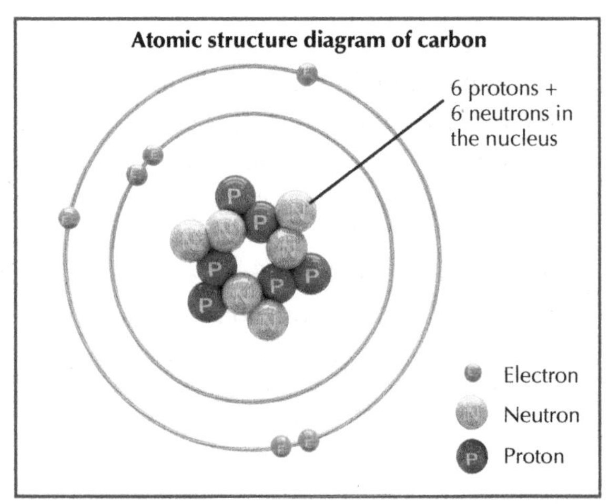

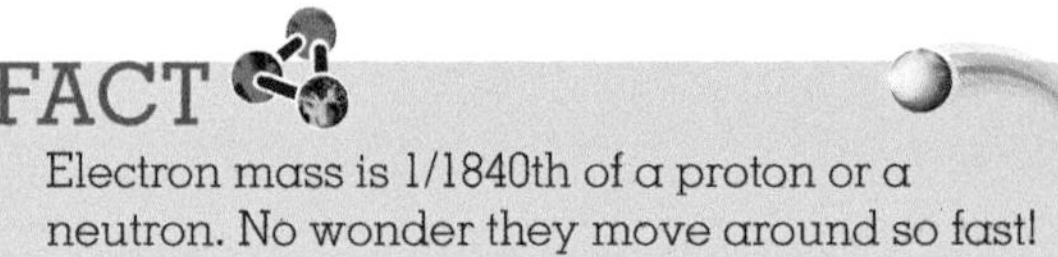

Electron mass is 1/1840th of a proton or a neutron. No wonder they move around so fast!

1.2: Electron arrangement

Complete the following table. (The first one has been done for you.)

Element/Ion symbol	Atomic number	Electron configuration
H	**1**	**1**
		2
Li		
	4	
B		
	6	
		2, 5
O		
	9	
		2, 8
	11	
Mg		
		2, 8, 3
	14	
	19	
Ca		
F^-		
Li^+		
	12	2, 8
	16	2, 8, 8

ISBN: 9780170260107

Ions

Ions form when atoms lose or gain electrons. They do this in order to gain **full valence** (outer) **shells** and so become more stable.

Metals lose electrons and become **positive cations**. Non-metals gain electrons and become **negative anions**.

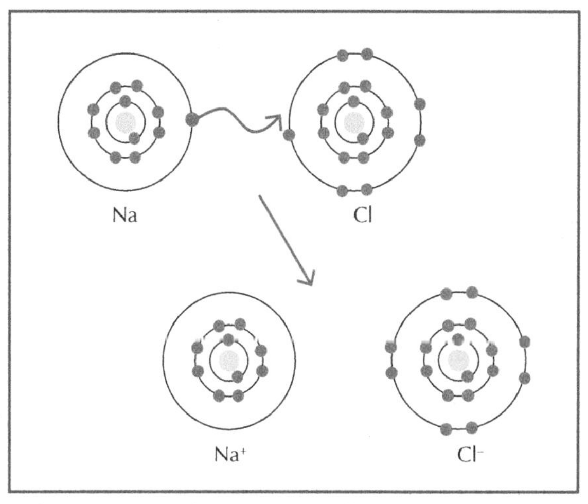

Sodium gives one of its electrons to chlorine making them both have full valence layers. The sodium having lost one electron has a +1 charge and the chlorine having gained an electron has a -1 charge.

FACT

The ammonium ion NH_4^+ has helped feed the world for nearly 100 years. First artificially made on an industrial scale in 1909, NH_4^+ has been used in the production of nitrogen-rich fertilisers that helped farm crops grow faster; the so-called 'green revolution'.

1.3: Ions

Fill in the following ion table to help you remember the ion formulas and names. (Note: there is no ion table given to you in Level 2.)

	Valency					
	1		2		3	
Metals	Lithium	Li^+	Magnesium			Al^{3+}
	Sodium	Na^+	Calcium		Iron (III)	
	Potassium		Copper (II)			
		Ag^+	Zinc			
		Cu^+	Iron (II)			
				Pb^{2+}		
				Ba^{2+}		
Non-metals	Fluoride		Oxide			
	Chloride		Sulfide			
	Bromide					
	Hydrogen					
Groups of atoms	Hydroxide		Carbonate		Phosphate	PO_4^{3-}
	Nitrate			SO_4^{2-}		
		NH_4^+				
		HCO_3^-				

ISBN: 9780170260107

What do the following key words mean?

Ion	
Cation	
Anion	
Valency	

Ionic bonding – putting ions together

When ions come together to form bonds they do so because it balances their charge. A positive cation is attracted to a negative anion because it has the opposite charge (opposites attract). This form of attraction between particles with opposite charges is called **electrostatic attraction**. Electrostatic attraction means the bond between two electrically charged particles, where one is positive and the other is negative.

Ions come together in specific amounts to balance out their charges and the result is neutral.

Example: one iron (II) ion (Fe^{2+}) will bond with two chloride ions (Cl^{-}), as the iron has two positive charges and so two negative charges are required to balance out the positive charges. This gives us the chemical formula $FeCl_2$. The name for this chemical formula is iron (II) chloride, which tells us that we have iron (II) (rather than iron (III) ions) and chloride present.

FACT 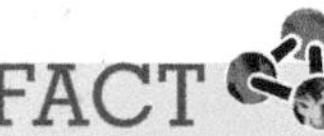

In 1884 Svante Arrhenius came up with the notion of ions. The word ion means 'to go' or 'to travel' in Greek, as ions move when conducting a current.

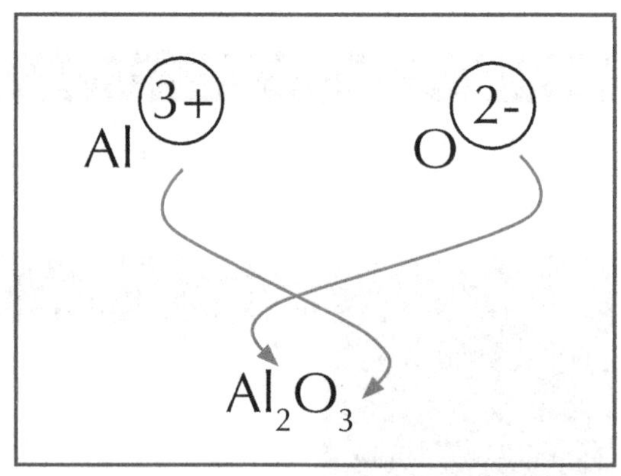

This diagram shows a commonly used method for writing ionic formulas. It is called 'the swap and drop method'. In this method you swap the numbers on the ion charges and subscript them next to the letters to give the exact ionic formula.

CHEMISTRY APPS

Table salt, or sodium chloride NaCl, is often extracted from sea water by letting the water evaporate. It can also be made by putting solid sodium and chlorine gas together; which gives a violent reaction where the metal burns in the green gas until you are left with a white solid of sodium chloride, NaCl. The atoms of sodium and chlorine have traded an electron and become Na^{+} and Cl^{-}, and are attracted to each other by an electrostatic attraction.

Reaction of sodium with water.

Sea salt drying.

 ISBN: 9780170260107

1.4: Ionic formulas

Try balancing the following ions together to find their chemical formula.

Ions	Chloride Cl^-	Hydroxide OH^-	Nitrate NO_3^-	Sulfate SO_4^{2-}	Sulfide S^{2-}	Carbonate CO_3^{2-}	Phosphate PO_4^{3-}
Sodium Na^+							
Ammonium NH_4^+							
Potassium K^+							
Calcium Ca^{2+}							
Magnesium Mg^{2+}							
Aluminium Al^{3+}							
Iron (II) Fe^{2+}							
Iron (III) Fe^{3+}							
Lead Pb^{2+}							
Copper (I) Cu^+							
Copper (II) Cu^{2+}							

What do the following key words mean?

Electrostatic attraction	
Ionic bond	
Neutral	

ISBN: 9780170260107

Metallic bonding

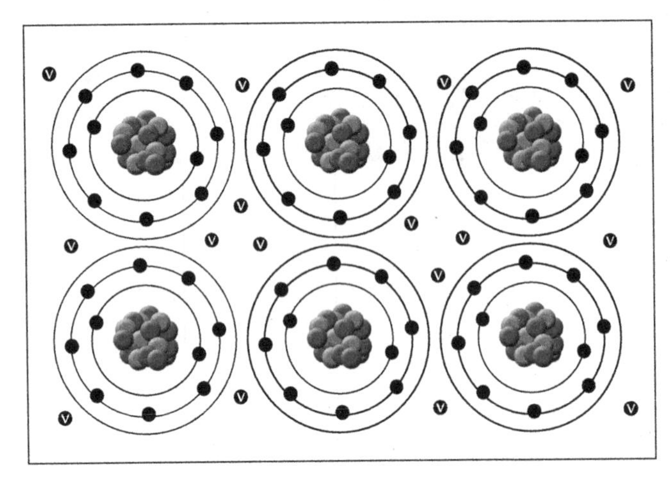

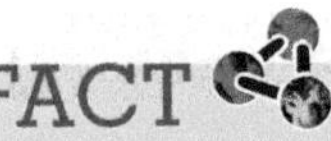

FACT

The most expensive metal is rhodium, which sells for over $200,000 (NZD) per kilogram. Platinum and gold, numbers two and three, trail far behind.

Metals have a giant structure made from metal nuclei and their delocalised valence electrons. **Delocalised** means the valence electrons in a metal are free to move – they have no specific locus or place. Metals are also said to have an electrostatic attraction, however, their attraction is between the positive nuclei and the negative electrons.

1.5: Ionic and metallic bonding

What are the similarities and differences between the bonds in a metal and the bonds in an ionic solid?

	Ionic bonding	Metallic bonding
Type of structure? Giant or Molecular	Giant	
Type of attraction		
Type of particles present in the structure – ions, atoms or molecules		Atoms
Type of bond	Ionic	
Diagram of bond		
Examples of where this type of bonding occurs	NaCl, $Mg(OH)_2$, $Al_2(CO_3)_3$	Mg, Al, Zn

ISBN: 9780170260107

CHECKPOINT 1
WHAT HAVE YOU LEARNED SO FAR?

Atoms, ions and metals

1 Explain the difference between a **compound** and a **mixture**.

2 Explain the difference between an **ion** and a **molecule**.

3 Write the electron configurations for the following atoms and ions:

a B

b Al

c F^-

d Mg^{2+}

4 Write the name of the following ionic compounds:

a CuF

b $Fe(OH)_3$

c $MgCO_3$

d Al_2O_3

5 Write the ionic formula for the following ionic compounds:

a Sodium chloride

b Lead carbonate

c Zinc phosphate

d Iron (II) hydrogen carbonate

6 Explain the term **ionic bond**. Your answer should include: what it looks like and the type of attraction between the particles.

7 Explain the term **metallic bond**. Your answer should include: what the bond looks like and the type of attraction between the particles.

ISBN: 9780170260107

Covalent bonding

Covalent bonding occurs when two or more non-metal atoms bond. Instead of giving or taking electrons, electrons are shared between atoms in order for the atoms to gain a full valence layer like the noble gases. There is an electrostatic attraction here between the electrons in the bond and protons in the nuclei of the atoms.

The simplest example is hydrogen or H_2, each hydrogen atom in the bond shares its one valence electron in order for them both to have two.

We can show this by drawing the electrons as dots in what is known as an **electron dot diagram** or **Lewis diagram**. In a single covalent bond there are always two electrons, represented as two dots or a line in a Lewis diagram.

H• + •H ⟶ H:H or H–H

Lewis structures of covalent molecules

We can draw Lewis diagrams for other molecules with different atoms present using the following rules shown in these examples:

Example 1: **Hydrogen fluoride, HF**

1. Add up all the valence electrons for the atoms in the molecule.
   HF: H has 1 valence electron and F has 7.
   Total electrons in HF = 7 + 1 = 8
   Divide by 2 into the number of pairs. 8/2 = 4 pairs of electrons.
2. The atom with the most incomplete octet is the central atom.
   In HF's case both are missing one electron. However as there are only two atoms we need not worry about this step at this stage.
3. Create single bonds first between each atom and the central atom. (A single bond remember is drawn as two dots or a line.)

   H – F or H : F
4. Add electrons in pairs around each atom until all atoms have a full valence shell*.

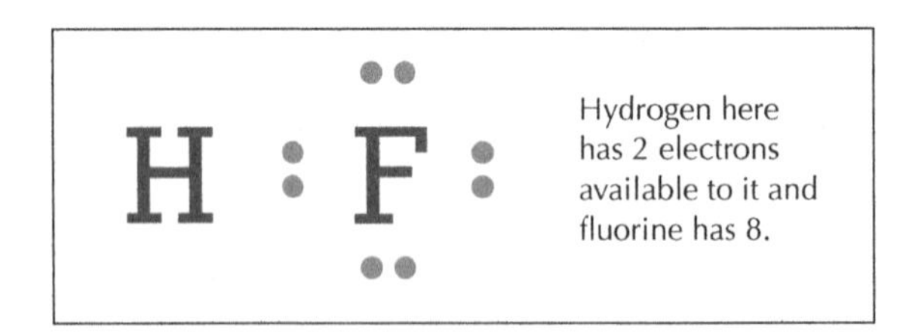

Hydrogen here has 2 electrons available to it and fluorine has 8.

5. Finally add up all the dots and check it adds up to the total number you calculated in Step 1. There are 8 dots in HF which add up to the 8 valence electrons we had.
6. If all atoms have their maximum number of electrons surrounding them and Step 5 works, then you are done. However sometimes we will need to add in double (4 electrons) or triple (6 electrons) bonds to make sure all atoms have full valence shells.

* *Note most tend to have 8, however there are three exceptions: H has 2, Be has 4 and B has 6.*

ISBN: 9780170260107

Example 2: Carbon dioxide, CO_2

1 Carbon has 4 valence electrons and each oxygen has 6. Total electrons = 4 + 6 x 2 = 16

2 Carbon is the central atom as it needs four electrons to gain a full outer shell.

O C O

3 O – C – O or O : C : O

4

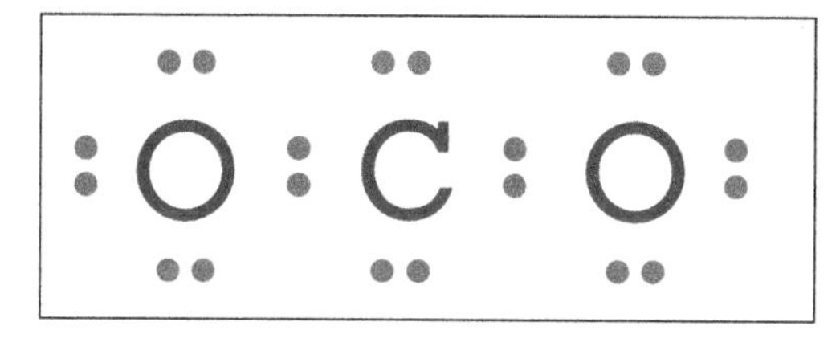

5 There are 20 dots but the molecule only has 16 electrons.

6 As the total number of dots does not equal the number of electrons we have to add in double bonds between the oxygens and the carbon, in order to ensure that all atoms have 8 electrons surrounding them and the number of dots adds up to 16.

:O::C::O: or :Ö = C = Ö:

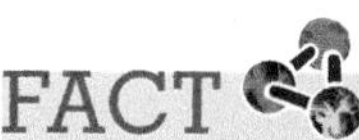

FACT

In 1916, Gilbert Newton Lewis, an American chemist, suggested that molecules were formed when atoms shared pairs of valence electrons. He was the first to show this in a diagram.

EXAMINER'S TIP

Don't forget the lone pairs of electrons!

Example: When drawing HBr you must remember to draw all 8 valence electrons around the Br as well as the bonding pair between the H and Br.

Like this...	Not like this...
H – Br : (with lone pairs above and below Br)	H – Br

ISBN: 9780170260107

1.6: Lewis diagrams

Try drawing the Lewis diagrams for the following molecules.

1 F_2	
2 HCl	
3 CO	
4 NH_3	
5 CH_4	
6 CF_4	
7 CH_3Cl	
8 BF_3 Don't forget B is an exception!	

ISBN: 9780170260107

Shapes of molecules

Using the Lewis diagrams on the previous page and VSEPR (Valence Shell Electron Pair Repulsion) theory we can work out the shape of a molecule in three dimensions.

VSEPR has two main points to help determine a shape:

1 Electrons repel each other to be as far apart as possible.
2 Lone pairs of electrons (those are the electrons that are not used in bonding) repel more strongly than bonding electrons.

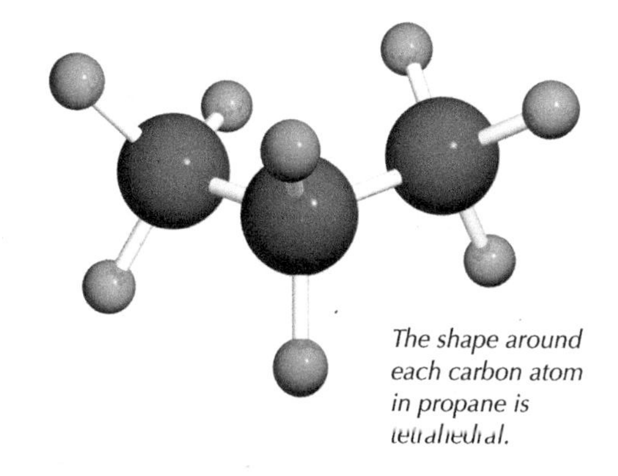

The shape around each carbon atom in propane is tetrahedral.

Number of bonding pairs	Number of lone pairs	Shape	Bond angle	Examples
1	0	Linear H – H	180°	H_2
2	2 (or 1)	Bent/V-shaped (diagram: O with two lone pairs bonded to H and H)	<109°	H_2O
3	0	Trigonal planar (diagram: B bonded to three F, 120°)	120°	BF_3
3	1	Trigonal pyramid (diagram: N with lone pair bonded to three H)	<109°	NH_3
4	0	Tetrahedral (diagram: C bonded to four H)	109°	CH_4

Note: When determining the shape of a molecule it is always best to start with a Lewis diagram in order to know how many bonding and lone pairs of electrons there are.

The shape diagrams above show how the atoms in a molecule are arranged in space.

KEY:

- (line) a bond in the plane of the page
- (wedge) a bond coming out of the plane of the page
- (hashed) a bond going into the plane of the page

EXAMINER'S TIP

1 Shapes form because the electron pairs repel each other as far apart as possible, not the atoms!
Example: in ammonia, NH_3 there are **four** regions of negative charge (1 lone pair and 3 bonding pairs) about the central atom N.

2 A multiple bond is counted as one region of negative charge.
Example: in carbon dioxide, CO_2, even though there are two double bonds (so four electrons connecting C to each O), we say there are **two** regions of negative charge (2 double bonds) about the central atom C.

1.7: Shapes of molecules

Draw the Lewis diagram, determine the shape, angle and draw the shape diagram for the following molecules.

Note: To help make shapes more easy to understand you could use some plasticine (as the atoms) and toothpicks (as the bonds) in order to see the following shapes in 3D.

Molecule	Lewis diagram	Shape and angle	Shape diagram
N_2			
CO_2			
H_2S			
SO_3			
CCl_4			
CH_3Cl			

ISBN: 9780170260107

CHEMISTRY APPS

Without molecules made up of non-metal atoms the surface of the Earth would look much more like the surface of the moon with no atmosphere, no water and therefore no life.

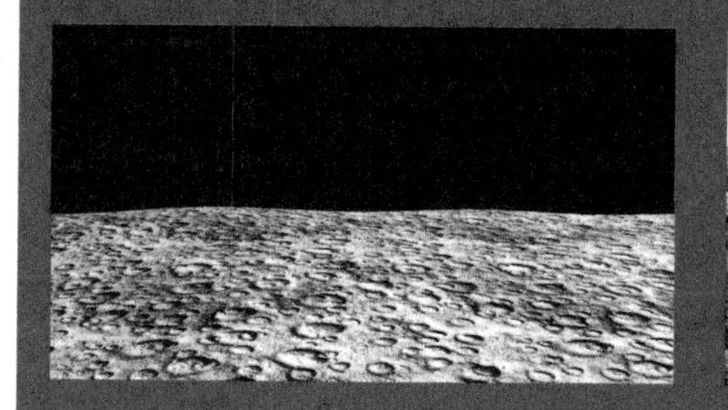

1.8: Explaining shapes of molecules

Below are examples of common types of questions in NCEA where you must explain why certain molecules have the shape that they do. The first one has been done for you to help you see what is expected in your answer.

Explain why H_2CO is trigonal planar and NCl_3 is trigonal pyramid.

H_2CO has three regions of negative charge about the central atom, two single bonds and one double bond. These repel each other as far apart as possible to the corners of a trigonal planar shape giving a bond angle of 120°.

As it has only three regions of negative charge it has a different shape and bond angle to NCl_3 which has four regions of negative charge.

NCl_3 has four regions of negative charge about the central atom, three single bonds and a lone pair of electrons. These repel each other to the corners of a tetrahedral shape giving a bond angle of slightly less than 109° (due to the extra repulsion from the lone pair). This gives it its trigonal pyramid shape.

1 Explain why NH_3 is trigonal pyramid and BF_3 is trigonal planar. In your answer you should include:

- A Lewis diagram for each molecule.
- How many lone pairs and bonded pairs of electrons (or regions of negative charge) there are.
- The bond angles present in each.
- Why their shapes are different.

ISBN: 9780170260107

2 Explain why H_2O is bent and CO_2 is linear. In your answer you should include:

- A Lewis diagram for each molecule.
- How many lone pairs and bonded pairs of electrons (or regions of negative charge).
- The bond angles present in each.
- Why their shapes are different.

Polar or non-polar

Covalent bonds involve the sharing of electrons. Unless the atoms present are identical there will be an unequal sharing of the electrons in a bond. We call this type of covalent bond **polar** as a partial charge separation occurs due to the unequal sharing of electrons.

In an ionic bond this charge separation is in full so that one atom becomes positive and another becomes negative. In polar bonds the charge separation is slight, so we use the symbol delta, δ to show this slight charge separation.

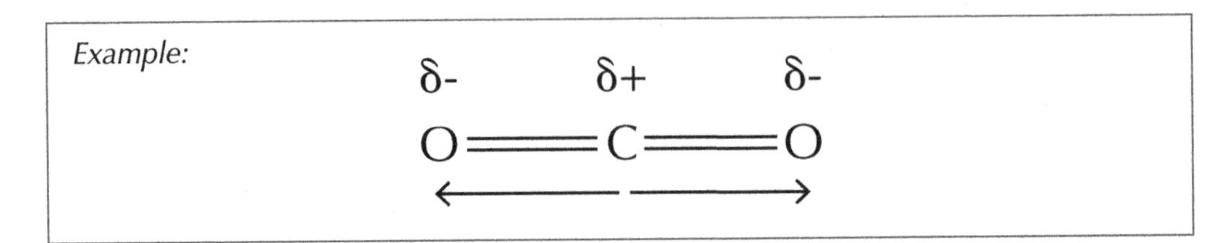

ISBN: 9780170260107

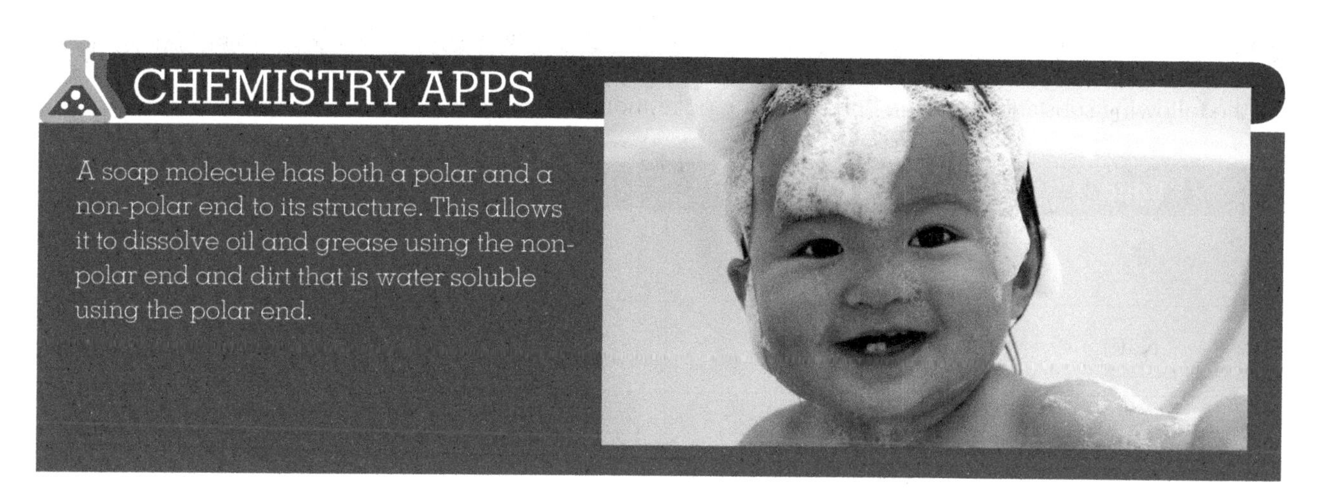

CHEMISTRY APPS
A soap molecule has both a polar and a non-polar end to its structure. This allows it to dissolve oil and grease using the non-polar end and dirt that is water soluble using the polar end.

Electronegativity

We determine which atom in the bond will become δ- and which will be δ+ by looking at its **electronegativity number**. Electronegativity is a value given to each atom to describe how attractive that atom is to electrons. You will not be required to remember the exact values but you should know in general that if there is a difference in electronegativity then there is a polar bond. If that difference is large enough then a full charge separation will occur. Total charge separation is one way of describing an **ionic bond**.

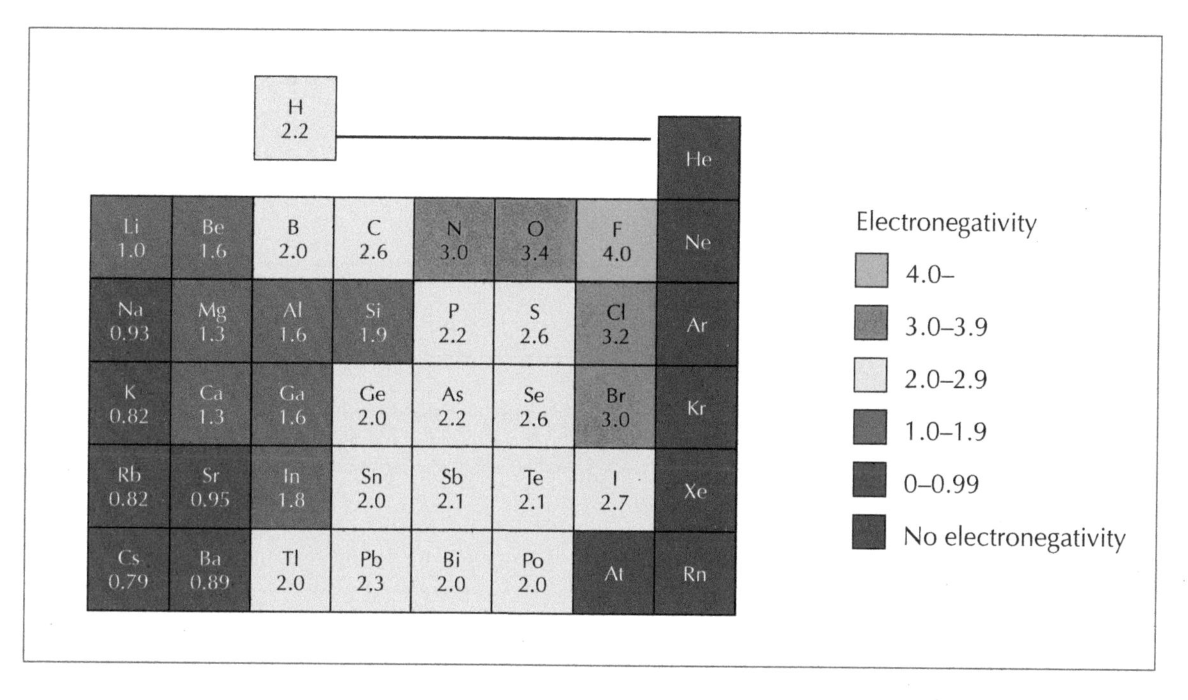

In general, electronegativity values increase across a period and decrease down a group of the periodic table. So, we can look at bonds as being on a continuum with non-polar covalent at one end where no charge separation occurs, polar bonds along the middle (depending on the magnitude of the electronegativity difference) and ionic bonds at the other end where full charge separation has occurred.

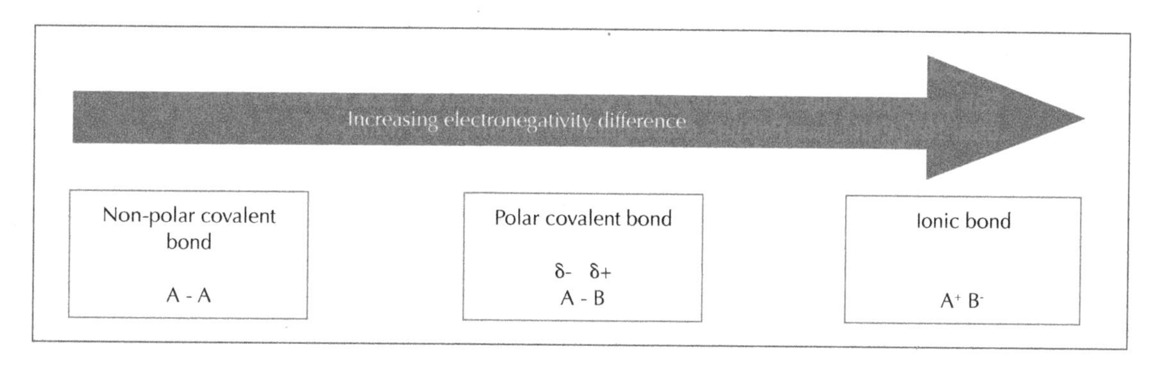

ISBN: 9780170260107

1.9: Polar, non-polar or ionic

Do the following substances contain bonds that are polar, non-polar or ionic?

Bond	Polar	Non-polar	Ionic
F_2	X	Yes	X
NaCl			
HBr			
O_2			
MgS			
CCl_4			
HCl			
NH_3			
Al_2O_3			
Fe_2S_3			

What do the following key words mean?

Covalent bond	
Non polar covalent bond	
Polar covalent bond	
Bond dipole	
Charge separation	
Electronegativity	

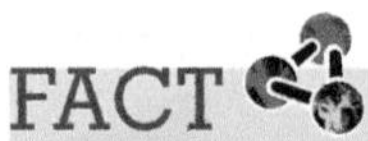

The concept of electronegativity has been around since 1809. However it wasn't until 1932 that Linus Pauling could quantify it (put it into numbers).

 ISBN: 9780170260107

Shape and polarity

Having polar bonds does not necessarily mean the molecule will be a polar molecule. This is because bond dipoles can be cancelled out if a molecule's shape is symmetrical. For example, carbon dioxide contains polar bonds but as its shape is symmetrical the bond dipoles cancel each other out.

$$\overset{\delta-}{O} = \overset{\delta+}{C} = \overset{\delta-}{O}$$

If a shape is asymmetrical like HCl, the molecule can be said to be polar as it contains polar bonds and the bond dipoles don't cancel.

$$\overset{\delta+}{H} - \overset{\delta-}{Cl}$$

1.10: Polar or non-polar molecules

Predict whether the following molecules are polar or non-polar. (The first one has been done for you.)

Molecule	Does it contain polar bonds?	Is the shape symmetrical?	Space to draw Lewis diagram if required	Is the molecule polar or non-polar?
O_2	No	Yes	$:\ddot{O}=\ddot{O}:$	Non-polar
NH_3				
CF_4				
CH_3Cl				
H_2O				

ISBN: 9780170260107

PCl_3				
SO_2				
SO_3				

EXAMINER'S TIP

A molecule may contain polar bonds but be non-polar overall. Don't confuse polar bonds and polar molecules!

Example: CBr_4 has polar bonds as there is a difference in electronegativity between the C and the Br, however as it has a symmetrical shape (tetrahedral) and all the bonds are the same (C – Br), the bond dipoles cancel.

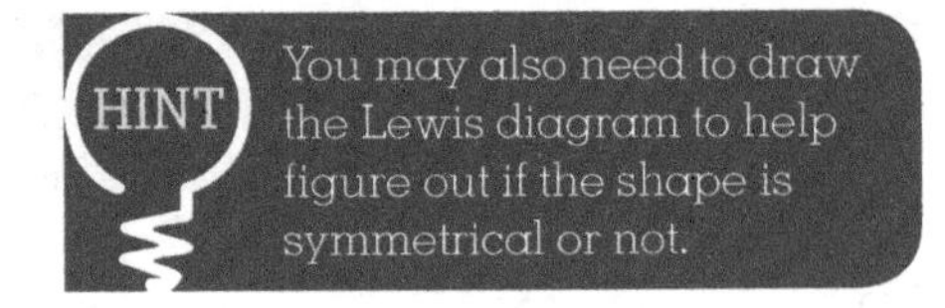

HINT
You may also need to draw the Lewis diagram to help figure out if the shape is symmetrical or not.

INVESTIGATION 1

Polar or non-polar substances

AIM: To find out which substances are polar using a charged balloon.

EQUIPMENT and CHEMICALS:

balloon	burette	nylon cloth	water
cyclohexane	ethanol	kerosene	beaker

SAFETY PRECAUTIONS:

Cyclohexane is flammable and kerosene is an irritant to eyes and inhalation.

METHOD:

1 Add around 20 mL of each liquid (water, cyclohexane, ethanol, kerosene) one at a time to the burette (rinse between uses).
2 Charge your balloon by rubbing it on nylon cloth.
3 Open the tap on the burette (make sure the beaker is underneath the stream!) and hold the balloon up close to the stream.

ISBN: 9780170260107

RESULTS

Substance	Observation	Polar or non-polar?
Water		
Cyclohexane		
Ethanol		
Kerosene		

QUESTIONS:

1 Why were certain substances attracted to the balloon? (Draw a picture to help illustrate your answer.)

2 What was it about the other substances that made them unattractive to the balloon?

CHECKPOINT 2
WHAT HAVE YOU LEARNED SO FAR?

Covalent bonds, shape and polarity

1 Explain the difference between the words **ionic bond** and **polar covalent bond**.

2 Explain the difference between the bond in NaF and in F_2.
In your answer you should:
- State what type of particle is found in each bond – atom, ion.
- State what happens to the electrons in each bond.

3 Fill in the blanks in the following table

Molecule	Lewis diagram	Shape name and diagram	Polar or non-polar molecule
CH_3F			
BH_3			
Cl_2O			
Br_2			

ISBN: 9780170260107

CHECKPOINT 2
WHAT HAVE YOU LEARNED SO FAR?

4 Both PH_3 and CH_4 have four regions of negative charge present in their molecules. Explain why they have different shapes, despite having the same number of regions of negative charge.

INVESTIGATION 2

Forces between particles

AIM: To relate the melting point of a substance to the forces between its particles.

EQUIPMENT and CHEMICALS:

sodium chloride, NaCl	sulfur, S_8	charcoal, C
iodine, I_2	potassium iodide, KI	ice, H_2O
silica, SiO_2	large tin lid	Bunsen burner

SAFETY PRECAUTIONS:

I_2 and S_8 are eye, skin and respiratory irritants.

INSTRUCTIONS:

- Your teacher will demonstrate heating iodine and sulfur (in a fume cupboard).
- You will heat the other substances listed in the table and record your observations for each – commenting on how easily they melt.

RESULTS:

Substance	Type of substance	Observations (ease of melting)
Iodine	Molecular	
Sulfur	Molecular	
Charcoal	Covalent network	
Sodium chloride	Ionic	
Silica	Covalent network	
Ice	Molecular	
Potassium iodide	Ionic	

QUESTIONS:

1 Which substances melted? What does their melting tell you about the force of attraction between their particles?

2 What happens to a substance when it melts?

 ISBN: 9780170260107

Properties of ionic solids

Ionic solids have the following properties that are related to their structure:

- **They are brittle** – this is because when struck, ions of the same charge line up causing repulsion and a split down a plane of the solid.

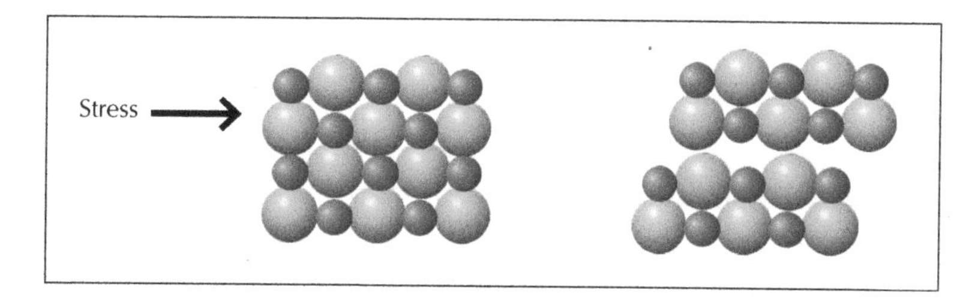

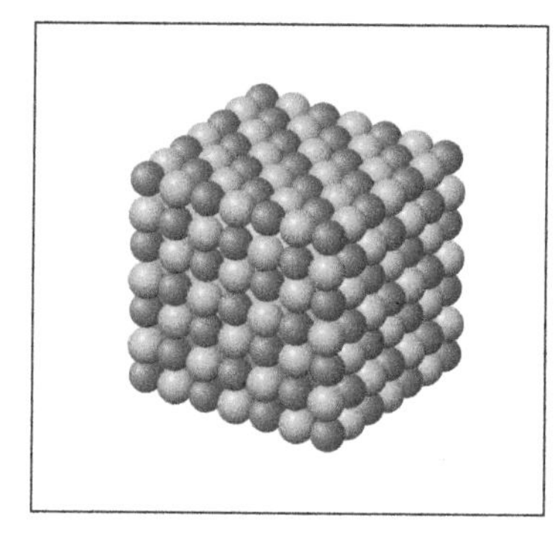

- **They have a high melting point** – this is because it takes a lot of energy to break apart the strong electrostatic forces of attraction between the ions.

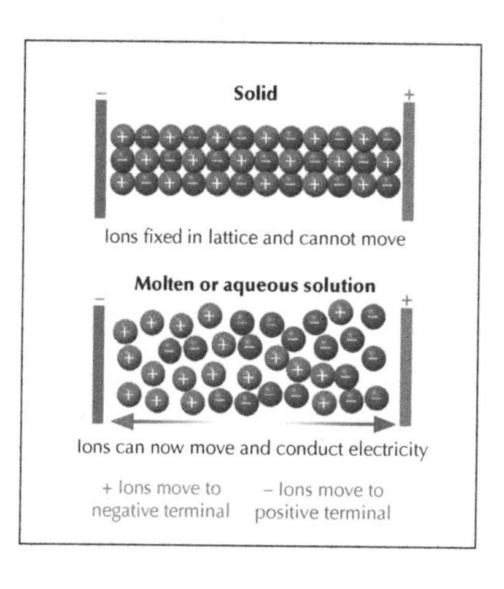

- **They conduct electricity when molten or dissolved in water (aqueous) but not when solid** – this is because when molten or aqueous, the ions are free to conduct a current, however, when solid the ions are in a fixed position and so are unable to carry a current.

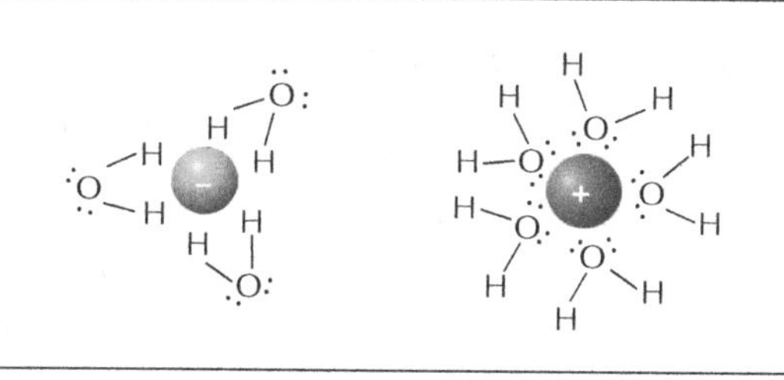

- **They dissolve in polar solvents like water** – this is because the force of attraction between the ions can be broken by the force of attraction between the ions and the water molecules.

EXAMINER'S TIP

For electricity to flow through any substance requires the substance to have mobile charged particles. These can be electrons but they can also be ions. Ionic compounds never have mobile electrons!

Example: in copper the valence electrons are free to move to carry a current, whilst the copper nuclei remain in place. In molten or aqueous copper sulfate the copper ions and the sulfate ions will conduct the current; there are no free moving electrons present.

CHEMISTRY APPS

Sports drinks contain electrolytes to help replace the sodium and potassium ions (and many more) our bodies need in order to keep being active. These electrolytes are just ionic compounds that have been dissolved in water and they are important for regulating water between the cells in our bodies as well as for transmitting nerve impulses from our brain.

ISBN: 9780170260107

Properties of metallic solids

Metallic solids have the following properties that are related to their structure:

- **They are malleable** – malleable means metals can be hammered or pressed into many different shapes without the bonds breaking. This is because the electrons are delocalised in a metal and so able to move around freely without breaking the electrostatic attraction between the metal nuclei and the electrons.

- **They are ductile** – ductile means they can be drawn into wires. This is for the same reason as above, due to their delocalised electrons.

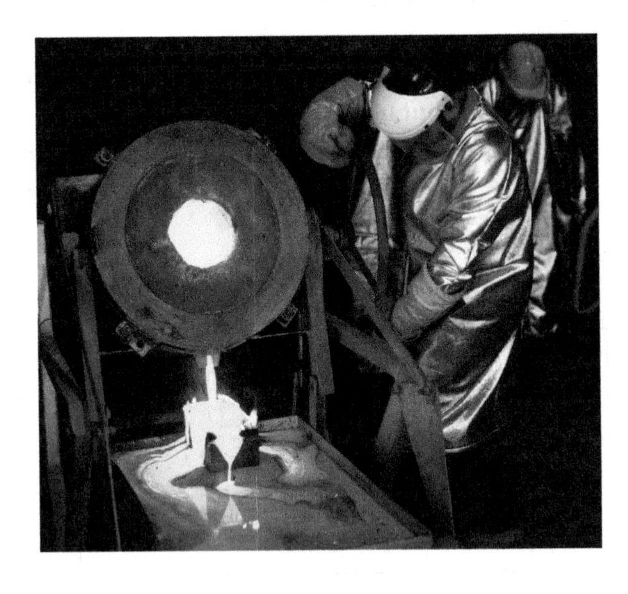

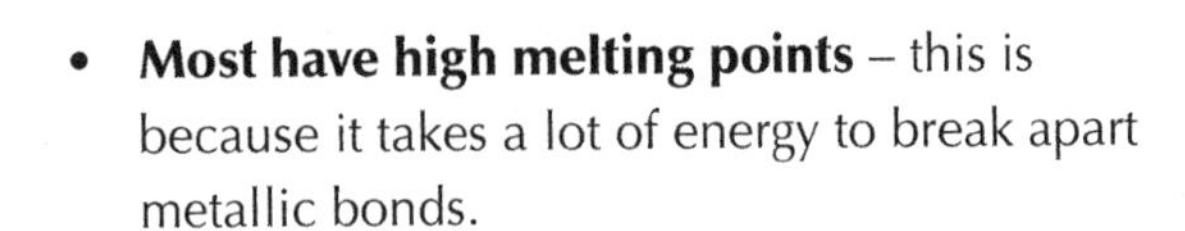

- **Most have high melting points** – this is because it takes a lot of energy to break apart metallic bonds.

- **They can conduct current when solid or liquid** – this is because the electrons are free to move throughout the solid or liquid to conduct a current.

- **They can conduct heat** – this is because the electrons are free to vibrate along the metal to transport the heat energy from one end to another via kinetic energy.

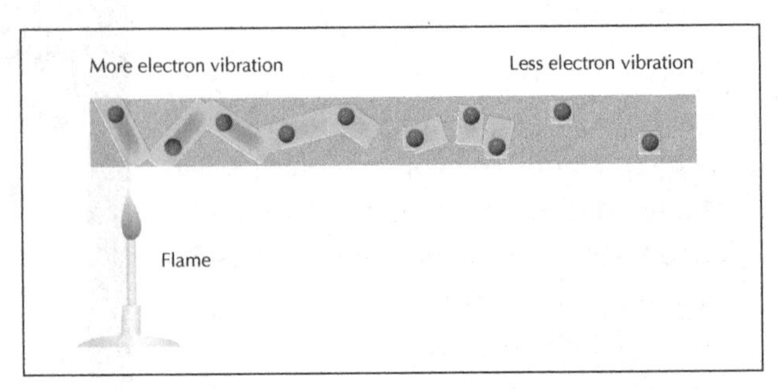

 ISBN: 9780170260107

What do the following key words mean?

Brittle	
Conduct	
Malleable	
Ductile	
Aqueous	

EXAMINER'S TIP

Always relate a physical property to the structure and bonding present in that substance.

Example: metals are malleable because the valence electrons are able to move freely throughout the solid without breaking their electrostatic attraction to the metal nuclei.

1.11: Properties of solids

1 Explain why the melting point for aluminium (660°) and aluminium sulfate (770°) are very similar in terms of their bonding and structure. In your answer you should include:

- The type of bonds found in each structure.
- Why they both have such high melting points.

2 Explain why aluminium conducts electricity when solid or molten but aluminium sulfate only conducts when molten or aqueous. In your answer you should include:

- The type of bonds found in each structure.
- Why they either conduct or do not conduct depending on their state.

ISBN: 9780170260107

3 Explain why aluminium is ductile but aluminium sulfate is brittle. In your answer you should include:

- The type of bonds found in each structure.
- Why aluminium is ductile but aluminium sulfate is brittle.

Properties of molecular solids

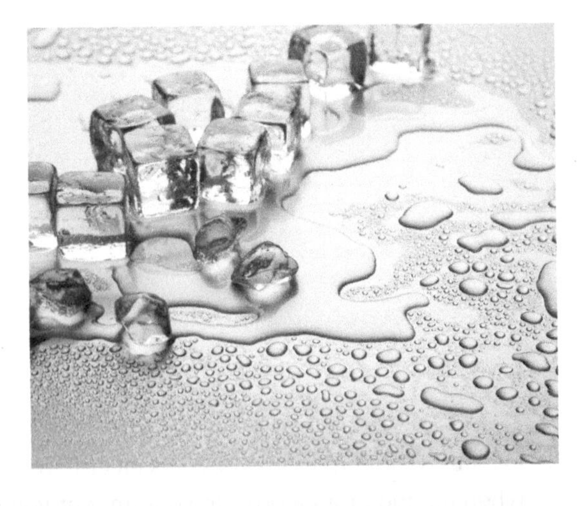

Molecular solids are solids made from molecules with either polar covalent or non-polar covalent bonds present. They have the following properties that are related to their structure:

- **They are soft** – the bonds between the molecules called intermolecular bonds require little energy to break.
- **They have low melting points and boiling points** – the bonds between the molecules, the intermolecular bonds, require little energy to break. (Note: the bonds inside the molecules like the bond between H and Br in HBr is a strong polar covalent bond which will not break when melting or boiling the molecule.)
- **They do not conduct electricity or heat** – this is because they have no free electrons or ions to conduct a current.
- **Like dissolves like** – molecules that are polar will only dissolve in a polar solvent (like water), molecules that are non-polar will only dissolve in non-polar solvents (like cyclohexane).

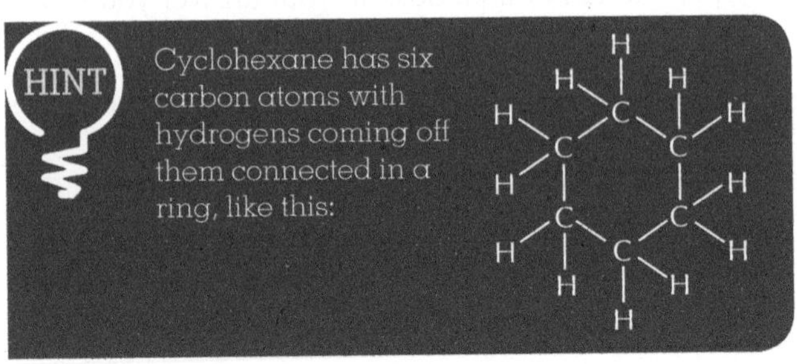

HINT
Cyclohexane has six carbon atoms with hydrogens coming off them connected in a ring, like this:
H
C

ISBN: 9780170260107

1.12: Properties of solids 2

1 Explain why oxygen is a gas at room temperature, but zinc and zinc oxide are solids. In your answer you should include:

- The type of bonding present in each.
- Why oxygen has a much lower melting point than Zn or ZnO.

2 Explain why sodium chloride dissolves in water but octane C_8H_{18} (the main component of petrol) does not. In your answer you should include:

- The type of bonding present in each.
- A diagram showing why the ions in sodium chloride are attracted to the water molecules.
- Why the type of bonds present in a structure determine its solubility in a solvent.

INVESTIGATION 3

Solubility and bonding

AIM: To relate the solubility of a substance to its bond type.

EQUIPMENT AND CHEMICALS:

sugar, $C_6H_{12}O_6$
copper sulfate, $CuSO_4$
silica, SiO_2
graphite, C
potassium iodide, KI
cyclohexane, C_6H_{12}
aluminium, Al
wax pellets
test tubes
iodine, I_2
ethanol, C_2H_5OH

SAFETY PRECAUTIONS:

Iodine is an irritant and cyclohexane is very flammable.

INSTRUCTIONS:

1 Test the substances listed below for solubility in water by placing one or two crystals / drops of each, in 2 mL of water and mixing thoroughly.
2 Repeat using cyclohexane instead of water. Record observations.

RESULTS:

Substance	Type of bonding	Observation in water	Observation in cyclohexane
Potassium iodide	Ionic		
Sugar			
Copper sulfate			
Silica			
Wax			
Ethanol			
Iodine			
Graphite			
Aluminium			

ISBN: 9780170260107

QUESTIONS:

1 Complete the following sentences:

Substances that dissolved in water were ______________________________

Substances that dissolved in cyclohexane were ______________________________

Insoluble substances in both water and cyclohexane were ______________________________

2 Why are ionic substances able to dissolve in water? (Draw a picture to help illustrate your answer.)

____

____

____

____

____

____

3 For the molecular substances that dissolve in water, what does this tell you about their overall polarity? How do you know?

____

____

____

____

____

____

____

____

____

____

____

____

____

ISBN: 9780170260107

INVESTIGATION 4

Making models of graphite and diamond

AIM: To build models of diamond and graphite in order to investigate the properties these substances possess.

Diamond

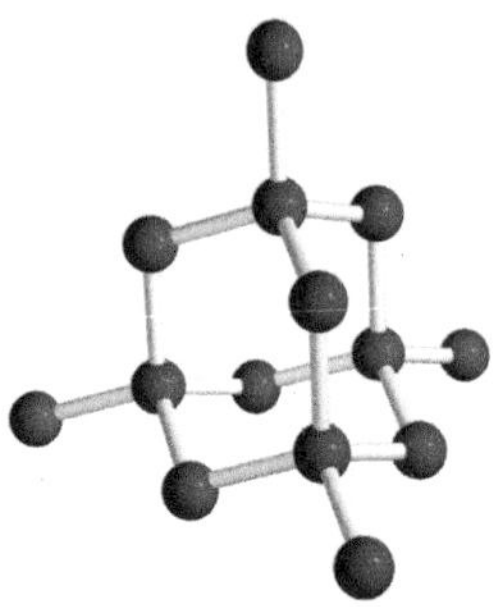

Graphite

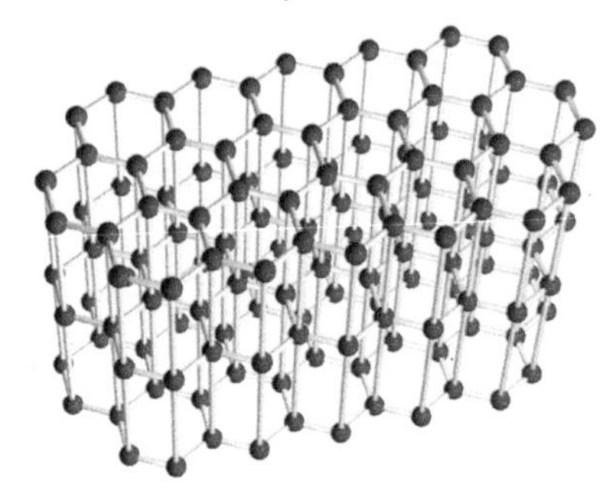

EQUIPMENT:

24 pieces of straw cut to 5 cm in length
plasticine
3 pieces of string about 15 cm in length

METHOD:

1. Divide the number of straws in half.
2. Use one half of the straws in a tetrahedral arrangement to represent diamond using the plasticine as the carbon atoms and the straws as the bonds between them.
3. Now use the other half to build two hexagons to represent graphite. Connect the two hexagons as shown in the diagram above using the pieces of string, so that in the centre of one hexagon is one of the atoms from the other hexagon.

QUESTIONS:

1. Which structure appeared to show more strength? Why?

____

____

____

____

____

____

2. Looking at the structures of each why do you think diamond is the hardest substance on Earth but graphite can be used as a lubricant as it is soft?

____

____

____

____

____

____

ISBN: 9780170260107

Properties of covalent network solids

These are solids which are generally hard and have some interesting properties. There are three common ones asked about in NCEA Level 2 examinations, however there are many more examples with unique properties.

1 Diamond

Diamond is made from covalently bonded carbon atoms with no intermolecular forces. Each carbon atom is connected to four other carbon atoms in a tetrahedral arrangement. This arrangement is very hard (diamond is the hardest-known substance on Earth) and requires very high heat to break apart. Diamond can not conduct a current as there are no free charged particles.

2 Graphite

Graphite is also made from covalently bonded carbon atoms, however each carbon atom is connected to three atoms allowing one free electron for each carbon. The carbon atoms are arranged in hexagonal rings and in layers. In between the layers are weak intermolecular bonds which keep the two-dimensional structure together. This means graphite is soft and makes an excellent lubricant as the layers can slide over the top of each other. Due to the free electron on each carbon graphite can also conduct a current.

3 Silicon dioxide

Silicon dioxide has silicon atoms which are covalently bonded to four oxygen atoms. It is a hard structure due to the presence of covalent bonds and lack of intermolecular bonds. We commonly know it as sand (note the yellow colour is an impurity). It also has a high melting point due to its strong covalent bonds.

What do the following key words mean?

Intermolecular bond	
Intramolecular bond	
Lubricant	
Tetrahedral	

CHEMISTRY APPS

In 2010 Andre Geim and Konstantin Novoselov won the Nobel Prize in Physics for their work with graphene. Graphene is single layers of graphite, Geim and Novoselov split the layers by using adhesive tape repeatedly sticking and unsticking it to get increasingly thinner pieces. Graphene has shown to be 200 times stronger than steel and may also be used as a conductor of electricity in the future.

ISBN: 9780170260107

1.13: Properties of covalent network solids

1 Explain why graphite conducts electricity but diamond does not.

2 Explain why silicon dioxide has a high melting point.

1.14: NCEA style questions – types of solids

A common question on this is given in the table on the next page. The first few have been done for you to help get you started.

1 The **substances** will always be given to you but read carefully as some common ones can trip students up, for example, silicon dioxide which is a covalent network solid as opposed to sulfur dioxide which is molecular.

2 The **type of particle** can only be *molecules* for molecular solids like iodine, *ions* for ionic compounds like sodium chloride, or *atoms* for metals like magnesium, or covalent network solids like diamond.

3 The **attractive force between particles** can only be *intermolecular* if it is a molecular solid, *ionic* if it is an ionic solid, *metallic* if it is a metallic solid, or *covalent* if it is a covalent network solid.

ISBN: 9780170260107

Substance	Type of particle	Attractive forces between particles
Iodine, I_2	Molecules	Intermolecular
Sodium chloride, NaCl	Ions	Ionic
Magnesium, Mg	Atoms	Metallic
Diamond, C	Atoms	Covalent
Zinc hydroxide, $Zn(OH)_2$		
Aluminium oxide, Al_2O_3		
Water, H_2O		
Oxygen, O_2		
Hydrochloric acid, HCl		
Octane, C_8H_{18}		
Zinc, Zn		
Graphite, C		
Silicon dioxide, SiO_2		
Lead, Pb		

CHECKPOINT 3

WHAT HAVE YOU LEARNED SO FAR?

Summary of the types of solids

Type of solid	Type of particle present in it – ions, atoms or molecules	Bonds present in the structure – ionic, metallic, covalent, intermolecular	Properties of the structure – hard/soft, conductor/insulator, high/low melting point	Diagram of the structure	Examples
Ionic					
Metallic					
Molecular					
Covalent network					

ISBN: 9780170260107

INVESTIGATION 5

Observing chemical reactions

AIM: To observe changes in heat in chemical reactions.

EQUIPMENT and CHEMICALS:

sodium hydroxide pellets NaOH
ammonium chloride NH_4Cl
test tubes
spatula
magnesium ribbon Mg
2 $molL^{-1}$ hydrochloric acid HCl
thermometer
anhydrous magnesium sulfate $MgSO_4$

SAFETY PRECAUTIONS:

NaOH and HCl are skin, eye and inhalation irritants and corrosive. NH_4Cl is an eye irritant.

METHOD and RESULTS:

For the following chemical reactions measure the temperature change that occurs for each.

1 Add 4–5 pellets of sodium hydroxide to 2 mL water in a test tube.

Temperature of the water before NaOH (°C)	
Temperature of the water with NaOH (°C)	
Overall change in temperature (°C)	

2 Add a spatula measure of solid ammonium chloride to 2 mL water in a test tube.

Temperature of the water before NH_4Cl (°C)	
Temperature of the water with NH_4Cl (°C)	
Overall change in temperature (°C)	

3 Add approximately 2 g of solid magnesium sulfate to 2 mL water in a test tube.

Temperature of the water before $MgSO_4$ (°C)	
Temperature of the water with $MgSO_4$ (°C)	
Overall change in temperature (°C)	

4 Add a 1 cm strip magnesium ribbon to 2 mL dilute hydrochloric acid in a test tube.

Temperature of the acid before Mg (°C)	
Temperature of the acid with Mg (°C)	
Overall change in temperature (°C)	

QUESTIONS:

1 Some chemical reactions release heat to their surroundings and are called exothermic reactions. Which of the reactions on the previous page are exothermic?

2 Other chemical reactions take heat from their surroundings and are called endothermic reactions. Which of the reactions on the previous page are endothermic?

3 Where do you think the heat comes from in exothermic reactions?

Exothermic and endothermic reactions

Reactions can be classified in terms of whether they release or absorb heat. **Enthalpy** (H) is a measure of how much heat a chemical system contains. **Change in enthalpy** (ΔH) is how much this heat content changes when a chemical reaction occurs. We classify this change in heat in two ways:

- **Exothermic reactions** release heat to their surroundings, in other words they get hot. For example, burning fuels. Their change in enthalpy (ΔH) is always a negative value as their products have less energy than their reactants. The making of bonds is exothermic. We can represent this change in heat on a graph like this:

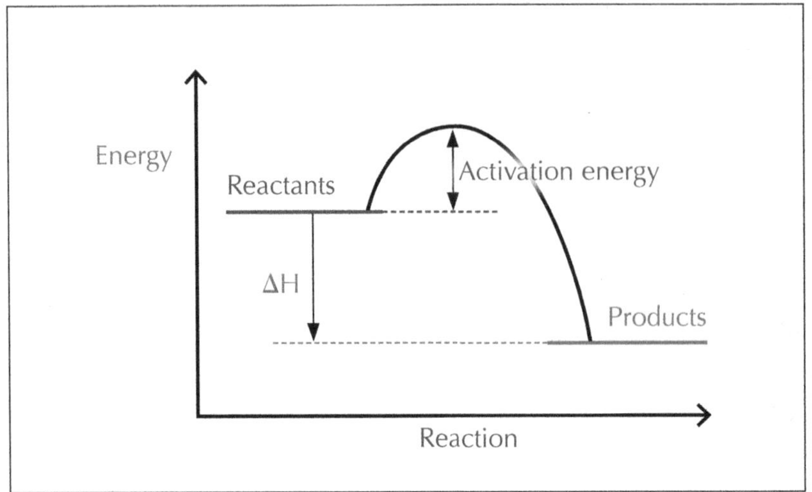

Note: Activation energy or E_a is the amount of energy required for a reaction to start.

ISBN: 9780170260107

- **Endothermic reactions** absorb heat from their surroundings, in other words they feel cold. For example, melting ice. Their ΔH is always positive as their products have more energy than their reactants. The breaking of bonds is an endothermic reaction. A graph showing an endothermic reaction looks like this:

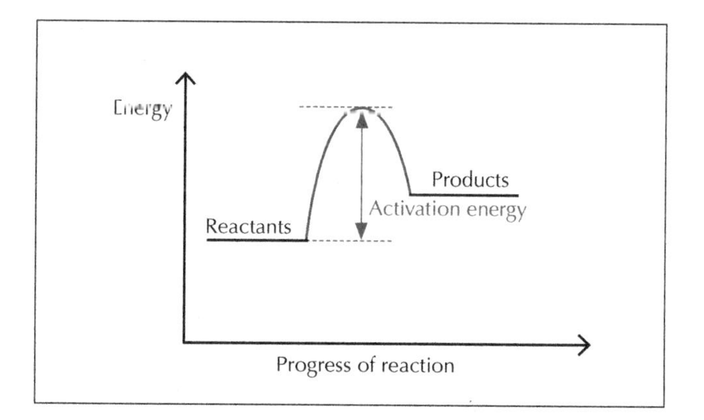

FACT

There are far fewer endothermic reactions than exothermic ones because endothermic reactions require such large amounts of activation energy.

CHEMISTRY APPS

Hand warmers are used to keep your hands warm in a cold situation. You may have already guessed that they work by creating an exothermic reaction however one common reaction they use is the reaction between oxygen and iron (an *oxidation reaction*) where iron becomes iron oxide. The other ingredients listed help catalyse the reaction and insulate in order to keep it warmer for a longer period of time.

1.15: Endothermic or exothermic reactions

Are the following reactions exothermic or endothermic? How do you know?

1 $2H_{2\,(g)} + O_{2\,(g)} \longrightarrow 2H_2O_{(g)}$ $\Delta H = -483.6\ kJmol^{-1}$

Exothermic Endothermic

Reason: ______________________________

____

____

____

2

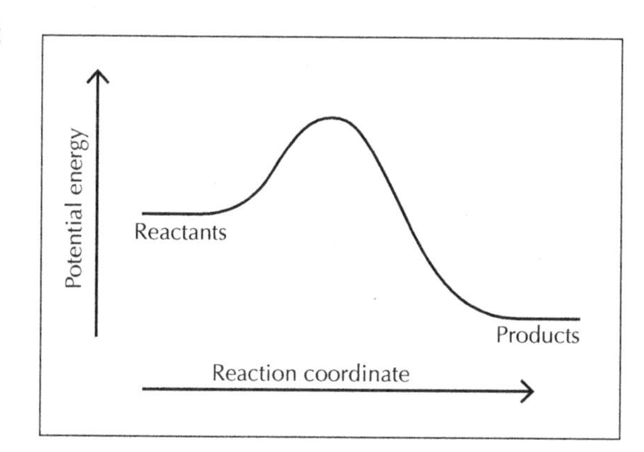

Exothermic Endothermic

Reason: ______________________________

____

____

____

____

3 The reaction of photosynthesis.

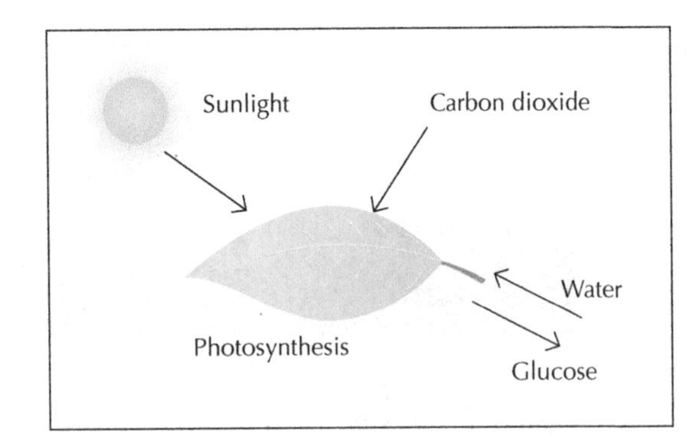

Exothermic Endothermic

Reason: ______________________________

____

____

____

____

Enthalpy

Enthalpy is the measure of the heat content of a substance. We can calculate the amount of heat released (exothermic) or absorbed (endothermic) for a reaction by multiplying ΔH by the number of moles of a reactant or product. This is because ΔH represents the amount of heat energy released in kilo Joules *per 1 mol of a reaction.*

Example: making methane

$C + 2H_2 \longrightarrow CH_4$ $\Delta H = -17.9\ kJmol^{-1}$

If we are given 2 mol of carbon then the total heat released will be:

Moles (n) x change in enthalpy ΔH = 2 mol x 17.9 $kJmol^{-1}$ = 35.8 kJ released

Note: We don't include the negative symbol for the ΔH in the calculation because you can't have negative energy but to show that the energy is released into the surroundings we write *released*. If it was an endothermic reaction we would write *absorbed*.

If we are given 2 g of carbon instead of a mol value we will need to work out how many moles in 2 g of C first:

$$\text{moles (n)} = \frac{\text{mass (g)}}{\text{molar mass (gmol}^{-1}\text{)}} \qquad \text{moles of C} = \frac{2}{12} = 0.167\ \text{mol}$$

Energy change = 0.167 mol x 17.9 $kJmol^{-1}$ = 2.98 kJ released

1.16: Enthalpy calculations

Calculate the maximum possible energy released or absorbed from the following reactions:

a 0.80 mol of hydrogen gas $2H_2 + O_2 \longrightarrow 2H_2O$ $\Delta H = -478\ kJmol^{-1}$

____

____

____

b 52 mol of carbon $2C + 3H_2 \longrightarrow C_2H_6$ $\Delta H = -20\ kJmol^{-1}$

____

____

____

 ISBN: 9780170260107

c 50g of cyclohexane, C_6H_{12} $C_6H_{12} + 9O_2 \longrightarrow 6H_2O + 6CO_2$ $\Delta H = -3920\ kJmol^{-1}$

d 12g of carbon dioxide, CO_2 $6CO_2 + 6H_2O \longrightarrow C_6H_{12}O_6 + 6O_2$ $\Delta H = +469\ kJmol^{-1}$

e 3 kg of propane, C_3H_8 $C_3H_8 + 5O_2 \longrightarrow 4H_2O + 3CO_2$ $\Delta H = -2220\ kJmol^{-1}$

Bond enthalpies

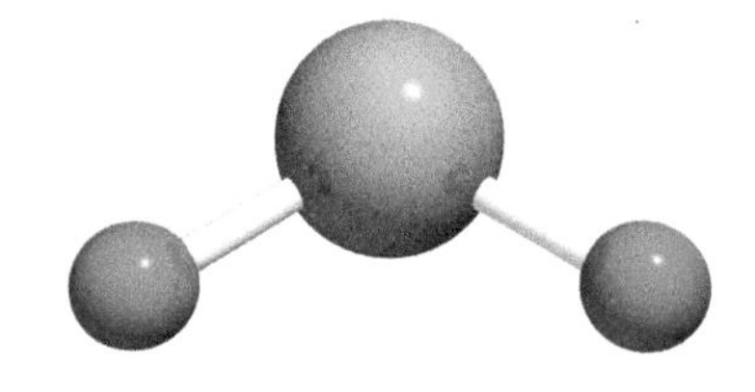

We can work out the energy required to break and form the bonds giving the overall change in enthalpy, ΔH, of a reaction using **bond enthalpy values**. A bond enthalpy value is the amount of energy required to break one mole of a chemical bond at room temperature (25 °C) and pressure (1 atmosphere). By adding up all the bonds broken and taking away all the bonds that are formed we can get the change in enthalpy (ΔH) of a reaction.

Example: making water

$2H_2 + O_2 \longrightarrow 2H_2O$

Bond type	Bond enthalpy ($kJmol^{-1}$)
H – H	432
O = O	494
O – H	459

Note: Bond enthalpy values are always positive or endothermic as it takes energy to break bonds.

1 First draw all the individual bonds in the molecules using Lewis structures.

2 H – H + Ö ═ Ö   2H–Ö–H

2 Add up the total value of all the **bonds broken** in the reactants.
Bonds broken = 2 H – H + O = O
= 2 x 432 + 494
= 1358 $kJmol^{-1}$

3 Add up the total value of all the **bonds formed** in the products.
Bonds formed = 4 O – H
= 4 x 459
= 1836 $kJmol^{-1}$

4 Work out the ΔH value by: ΔH = bonds broken – bonds formed

ΔH = 1358 – 1836
= -478 $kJmol^{-1}$

Note: Remember the minus sign tells us this is an exothermic reaction.

ISBN: 9780170260107

1.17: Bond energies

Bond type	Bond enthalpy ($kJmol^{-1}$)	Bond type	Bond enthalpy ($kJmol^{-1}$)
C – H	413	H – Cl	431
O = O	494	C – Cl	328
C = O	799	N ≡ N	941
O – H	459	N – H	391
H – H	432	I – I	151
Cl – Cl	242	H – I	?

How much energy is released or absorbed when the following reactions take place?

1 $CH_4 + 2O_2 \longrightarrow CO_2 + 2H_2O$

2 $H_2 + Cl_2 \longrightarrow 2HCl$

3 $CH_4 + Cl_2 \longrightarrow CH_3Cl + HCl$

4 $N_2 + 3H_2 \longrightarrow 2NH_3$

What is the bond energy value of the missing H – I bond?

5 $H_2 + I_2 \longrightarrow 2HI \qquad \Delta H = +26.5\ kJmol^{-1}$

FACT 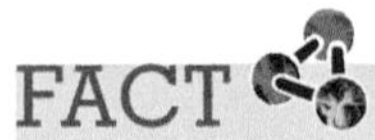

Bond enthalpies are an average as the value can change depending on what other atoms the atoms in the bond are connected to.

 ISBN: 9780170260107

EXAM-TYPE QUESTIONS

Before you get started here are some points to remember from the Examiner's Report:
- Don't use generalisations, use terms carefully.
- Plan your answer before you start.

Question one

a Draw the Lewis structures for each of the following molecules.

Molecule	Cl_2	SF_2	PBr_3
Lewis structure			
Molecule	SO_3	CF_4	F_2O
Lewis structure			

b Lewis structures for two molecules are given below.

Molecule	BeH_2	SO_2
Lewis structure	H : Be : H	:Ö : : S : Ö: (with lone pairs shown as dots above and below)

For each molecule, name the shape of the molecule and give a reason for your answer.

i BeH_2

Shape: ____________________

Reason: ____________________

____

____

ii SO_2

Shape: ____________________

Reason: ____________________

____

____

Question two

a Draw the Lewis structures, name the shape of and shape diagrams for each of the following molecules.

Molecule	O_2	NCl_3	CCl_3H
Lewis structure			
Shape			
Shape diagram			

b Both NCl_3 and CCl_3H have four regions of negative charge about their central atom. Explain why despite this they do not form the same shape but have similar bond angles and the same number of regions of negative charge.

ISBN: 9780170260107

EXAM-TYPE QUESTIONS

Question three

For the following molecules:

i Circle the correct phrase that describes whether the molecule has polar bonds or non-polar bonds.

a H_2O **polar bonds** **non-polar bonds**

b O_2 **polar bonds** **non-polar bonds**

ii Justify your answers to **a** and **b** above.

iii Circle the phrase that describes if the following molecules are polar or non-polar.

a CH_4 **polar molecule** **non-polar molecule**

b NH_3 **polar molecule** **non-polar molecule**

iv Justify your answers to **a** and **b** above.

ISBN: 9780170260107

Question four

a State the shape and draw the shape diagram of CCl_3H and CCl_4 below.

Molecule	Shape diagram
CCl_3H	Shape:

Molecule	Shape diagram
CCl_4	Shape:

b Circle the word that describes the **polarity** of each of the molecules $\mathbf{CCl_4}$ and $\mathbf{CCl_3H}$.

CCl_4	**polar**	**non-polar**
CCl_3H	**polar**	**non-polar**

For each molecule, justify your choice.

ISBN: 9780170260107

EXAM-TYPE QUESTIONS

Question five

a Complete the table below by stating the type of solid, the type of particle present, and the bonding (attractive forces) between particles in the **solid** state.

Solid	Type of solid	Type of particle present	Attractive forces between particles
Cl_2			
$SiCl_2$			
SiO_2			
Graphite			

b Discuss the conductivity and hardness of both graphite and silicon dioxide in terms of their bonding and structure.

ISBN: 9780170260107

Question six

a Complete the table below by stating the type of solid, the type of particle present, and the bonding (attractive forces) between particles in the **solid** state.

Solid	Type of solid	Type of particle present	Attractive forces between particles
Mg			
$MgCl_2$			

b Discuss the difference in conductivity and malleability of magnesium and magnesium chloride referring to their structure and bonding.

ISBN: 9780170260107

Question seven

a Discuss why cyclohexane, C_6H_6 has a melting point of 6.5 °C and sodium chloride has a melting point of 801 °C. Refer to their structure and bonding. Use the illustrations of cyclohexane below to help explain your answer.

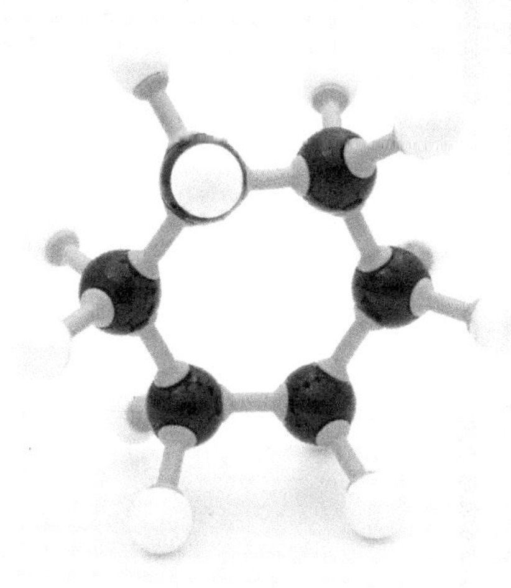

b Discuss why sodium chloride will dissolve in water but won't dissolve in cyclohexane. Refer to their structure and bonding. Draw a diagram to help support your answer.

Question eight

a PBr_3 and BH_3 both have three atoms attached to their central atom. Explain why they have different shapes, despite having the same number of atoms attached.

b Explain why PBr_3 is polar and BH_3 is non-polar. Refer to their shape and ability to have polar bonds.

 ISBN: 9780170260107

Question nine

The reaction of the formation of hydrochloric acid is shown below. Use this to help answer the questions that follow.

$$\frac{1}{2} Cl_2 + \frac{1}{2} H_2 \longrightarrow HCl \qquad \Delta H = - \text{ kJ mol}^{-1}$$

a Circle the phrase below that correctly describes the type of reaction for the formation of hydrochloric acid.

Exothermic Endothermic

Reason:

b Calculate the change in enthalpy for the reaction in kJmol^{-1} from the bond enthalpy data below.

Bond type	Bond enthalpy (kJmol^{-1})
Cl – Cl	242
H – H	432
H – Cl	431

c Calculate the heat released when 10 g of hydrochloric acid gas is formed.

Question ten

The reaction of the combustion of ethane C_2H_6 is shown below. Use this to help answer the questions that follow.

$$C_2H_6 + 3\ ½\ O_2 \longrightarrow 3H_2O + 2CO_2 \qquad \Delta H = -\ 1559.7\ kJmol^{-1}$$

a Circle the phrase below that correctly describes the type of reaction in the combustion of ethane.

Exothermic   Endothermic

Reason:

b Calculate the heat released if 6 kg of ethane is burnt.

c Calculate the bond enthalpy of the O – H bonds given the following bond enthalpy values.

Bond type	Bond enthalpy ($kJmol^{-1}$)
C – H	413
O = O	494
C = O	799
C – C	347
O – H	?

 ISBN: 9780170260107

2.5 Demonstrate understanding of the properties of selected organic compounds (91165)

Learning outcomes

Tick off when you have studied this idea in class and when you have revised over that section prior to your assessment.

	Learning Outcomes	In class	Revision
1	Define the following key terms – hydrocarbon, functional group, homologous, saturated and unsaturated.		
2	Name, draw, predict the properties and state the reactions of alkanes.		
3	Recognise, draw, explain and name structural isomers.		
4	Name, draw, predict the properties and state the reactions of alkenes.		
5	Recognise, draw, explain and name geometric isomers.		
6	Name, draw, predict the properties and state the reactions of alkynes.		
7	Name, draw, predict the properties and state the reactions of haloalkanes.		
8	Name, draw, predict the properties and state the reactions of alcohols.		
9	Name, draw, predict the properties and state the reactions of amines.		
10	Name, draw, predict the properties and state the reactions of carboxylic acids.		
11	Define and give examples of the following types of reaction, substitution, addition, oxidation, polymerisation and elimination.		

ISBN: 9780170260107

Pre-test – What do you know?

Section 1: What is organic?

1 Can you name any products that are made from, or are a part of, oil?

2 Circle the substances below which you think are organic substances.

skin cells	diamond	calcium carbonate
rubber	plastic	vitamin D
petrol	graphite	water

3 What feature was in common between the substances you circled?

4 What does the word **organic** mean to you?

Section 2: Types of organic substances

1 Name as many plastic products as you can, that you see around you now.

2 Name as many hydrocarbon products as you can, that you see around you now.

3 Name as many types of alcohol as you can.

ISBN: 9780170260107

What is organic chemistry?

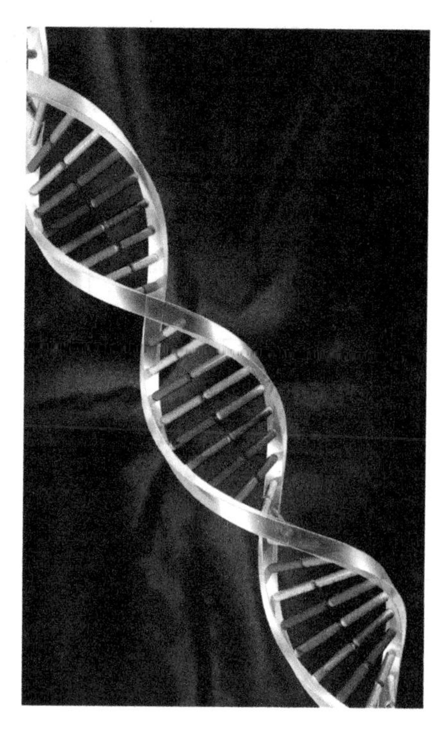

Originally an organic substance was any substance made by a living thing. Now we say it is the study of hydrocarbons (compounds made from carbon and hydrogen) and their derivatives. There are over 7 million organic compounds, but only 1.5 million inorganic ones.

Some organic substances are shown in the images below:

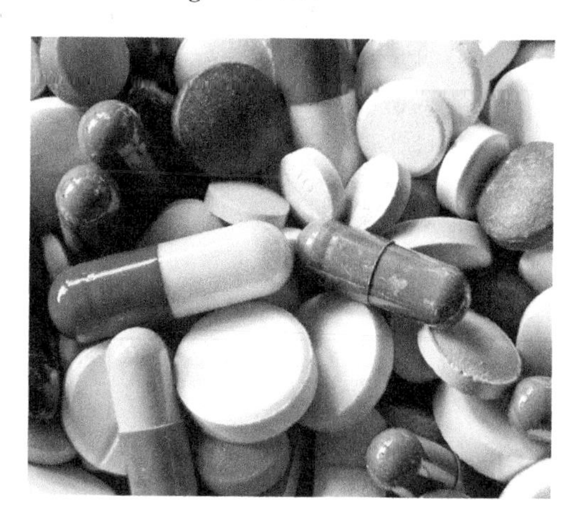

Pharmaceuticals, DNA, proteins, enzymes, pigments, oils, fats and plastics are all examples of organic compounds. To chemists, buying a piece of organically grown fruit is silly since all fruit is organic and a lot of the pesticides and herbicides sprayed on non-organic fruit are in fact organic compounds!

However along with the great things organic chemists have synthesised there are huge issues with waste and the depleting of oil reserves that we will have to overcome.

Organic compounds are grouped into families which have similar structures and properties. Each family has a distinctive **functional group**, a particular groups of atoms. **Homologous series** is a term often used instead of 'family'. You will have to be able to name a few of these functional groups, recall their properties and the way that they react. The groups you need to remember are recorded in the table on the following page.

ISBN: 9780170260107

Family and functional group	Structure	Examples
Alkanes C and H only, no functional group		Methane and butane, both of which are natural gases.
Alkenes C and H only, double bonds C = C		Ethene is the ripening agent given off by ripe bananas and other fruit, like apples.
Alkynes C and H only, triple bonds C ≡ C	$H-C\equiv C-H$	Ethyne is used for welding and cutting and is also known as acetylene.
Alcohols OH group(s)		Ethanol makes up hundreds of alcoholic beverages. Methanol is the main component of methylated spirits.
Haloalkanes Have atom(s) of F, Cl, Br or I		CFCs or chloro-fluoro carbons are a group of substances which were used in aerosols and refrigerators but have now been banned as they caused the hole in the ozone layer.
Amines Have NH_2 group(s)	H_2N-CH_3	Trimethylamine is the smell of old fish. When you add lemon juice you are neutralising the smell since amines are weak bases, and lemon juice contains acid.
Carboxylic acids Have COOH group(s)		Ethanoic acid is in vinegar and methanoic acid is an acid found in ants.

ISBN: 9780170260107

Why is carbon so special?

Carbon forms so many different compounds because it has four valence electrons and so it will usually form four covalent bonds. Below right, are different ways of showing the structure of methane.

Carbon can also do something quite unusual. It can form double and triple bonds with another carbon molecule as shown in the structure of ethene below:

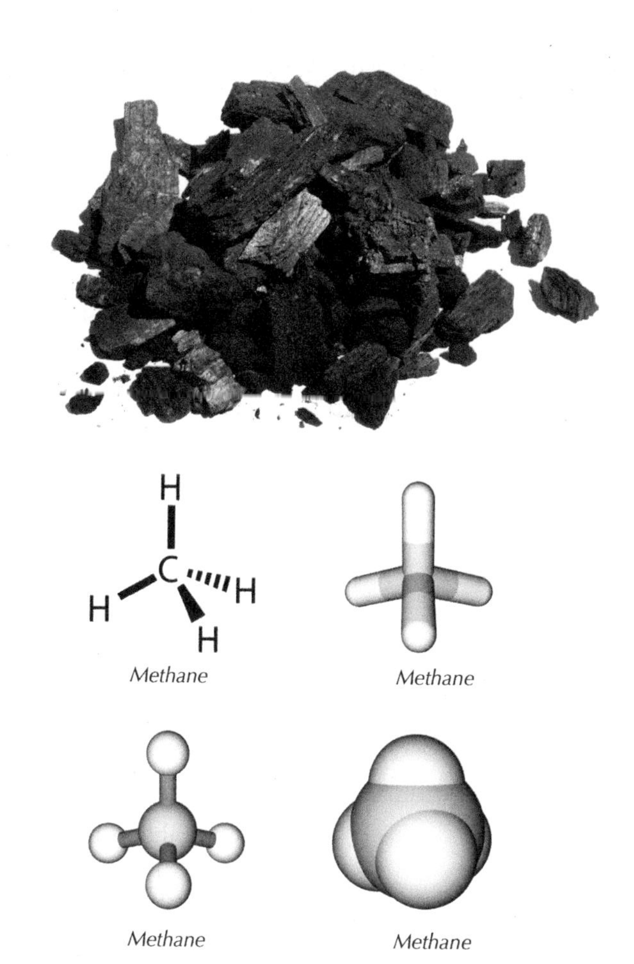

Methane *Methane*

Methane *Methane*

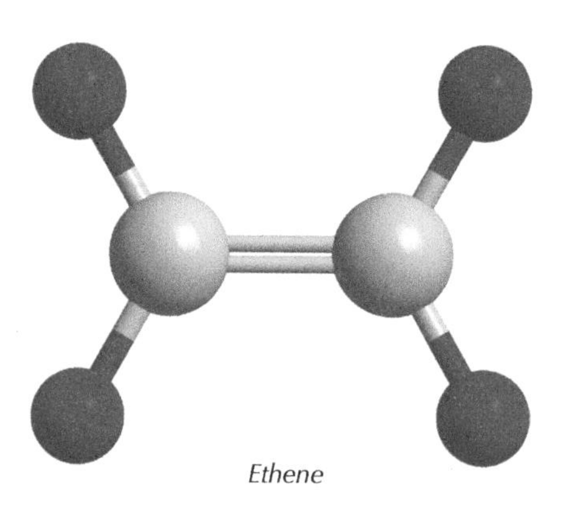

Ethene

There are two ways we draw the structure of organic molecules to show all the bonds they contain: structural formula (with every bond drawn out) and condensed structural formula (with only the atoms written in order of how the carbons are connected).

Structural formula	Condensed structural formula
H | H — C — H | H	CH_4
H H | | H — C — C — H | | H H	CH_3CH_3
H H H | | | H — C — C — C — O — H | | | H H H	$CH_3CH_2CH_2OH$
H H H O | | | // H — C — C — C — C | | | \\ H H H O — H	$CH_3CH_2CH_2COOH$ *(note the double bond connected to the O is not shown but since the carbon is only connected to two other atoms you need to know one of them has to be a double bond since carbons always have four bonds surrounding them)*

ISBN: 9780170260107

2.1: Structural and condensed structural formulas

Draw structural and condensed formula (make sure each carbon always has four bonds attached to it).

<table>
<tr><th></th><th>Structural formula</th><th>Condensed structural formula</th></tr>
<tr><td>1</td><td>H
|
H — C — O — H
|
H</td><td></td></tr>
<tr><td>2</td><td>H   H
\   /
C = C
/   \
H   H</td><td></td></tr>
<tr><td>3</td><td>H   H   H
| | |
H — C — C — C — H
| | |
H   H   H</td><td></td></tr>
<tr><td>4</td><td>H   H   H   H
| | | |
H — C = C — C — C — H
| |
H   H</td><td></td></tr>
<tr><td>5</td><td></td><td>CH_3Cl</td></tr>
<tr><td>6</td><td></td><td>CH_3CH_3</td></tr>
<tr><td>7</td><td></td><td>CH_2CHCH_3</td></tr>
<tr><td>8</td><td></td><td>CH_2CH_2</td></tr>
</table>

EXAMINER'S TIP

Make sure you draw out all the bonds and put in all the hydrogens when you draw structural formulas. Get into the habit now so you make sure you do it in the exam.

ISBN: 9780170260107

Naming organic compounds

The *first* step in naming an organic substance is stating how many carbon atoms are contained in one molecule. For that task we have special prefixes for each number:

Meth = 1 carbon
Eth = 2 carbons
Prop = 3 carbons
But = 4 C
Pent = 5 C
Hex = 6 C
Hept = 7 C
Oct = 8 C

2.2: Labelling the number of carbon atoms

Label the number of carbon atoms in the chain of the following organic substances using the prefixes shown above, the first two have been done for you:

	Organic substance	Prefix
1	$CH_3CH_2CH_3$	Prop
2	CH_3NH_2	Meth
3	CH_3COOH	
4	$CH_2{=}CH_2$	
5	CH_4	
6	$CH_3CH_2CH_2CH_2OH$	
7	$CH_3CH_2CH_2CH_2CH_2CH_2CH_2Cl$	
8	$CH_3CH_2CH_2CH_3$	

The *next* step is writing down which functional group an organic substance belongs to. This is usually given as a suffix or ending to the name. However haloalkanes (compounds containing halogens – fluorine, chlorine, bromine or iodine) and sometimes amines have it in front of their name before the number of carbons is stated.

ISBN: 9780170260107

Functional group	Atoms in functional group	Naming suffix	Examples
Alkanes	C – C	- ane	Methane, ethane, propane
Alkenes	C = C	- ene	Ethene, propene, butene
Alkynes	C ≡ C	- yne	Ethyne, propyne, butyne
Alcohols	C – O – H	- ol	Methanol, ethanol, propanol
Haloalkanes	C – Cl or C – F	Fluoro, chloro, bromo, iodo (note these are at the start of the name)	Fluoromethane, chloroethane, bromopropane
Amines	C – NH_2	- amine or amino (which is used at the start of the name)	Methanamine or aminomethane
Carboxylic acids	C = O, \ OH	- oic acid	Methanoic acid, ethanoic acid, propanoic acid

2.3: Labelling the functional group

Label the functional group and the number of carbons in the organic substances below

	Organic substance	Prefix and suffix
1	$CH_3CH_2CH_3$	Propane
2	CH_3NH_2	Aminomethane or methanamine
3	CH_3COOH	
4	$CH_2{=}CH_2$	
5	CH_4	
6	$CH_3CH_2CH_2CH_2OH$	
7	$CH_3CH_2CH_2CH_2CH_2CH_2CH_2Cl$	
8	$CH_3CH_2CH_2CH_3$	

The next step involves numbering what carbon the functional group is on, as each carbon in a chain can be a different compound if the functional group is on it. Alkanes do not need a number as they don't have a specific group attached to them like the others; they just have all carbon single bonded to other carbons or hydrogen atoms.

To number a carbon chain you start at the end that is closest to the functional group, as the functional group should get the lowest possible number.

For example:
We could number the following alkene in two ways …

$CH_3CH_2CH_2CH_2CH = CH_2$
1 2 3 4 5 6

$CH_3CH_2CH_2CH_2CH = CH_2$
6 5 4 3 2 1

ISBN: 9780170260107

The first numbering would give us the name hex-5-ene and the other would give us hex-1-ene. We always pick the name that gives the lowest number possible for the place of the functional group. You can also see we label them 1 and 5 rather than 2 and 6 as we pick the lowest number of carbon attached to the functional groups.

Functional group	Compound structure	Name	Notes
Alkenes	$CH_2 = CHCH_2CH_3$ 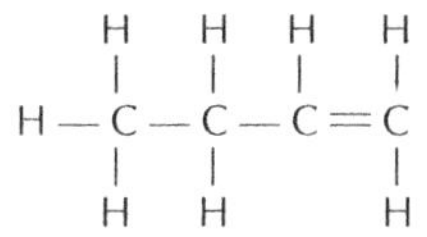	But-1-ene	We don't bother to number the position of the double bond in propene as either way you look at it the double bond will always be attached to carbon number one.
	$CH_3CH{=}CHCH_3$ 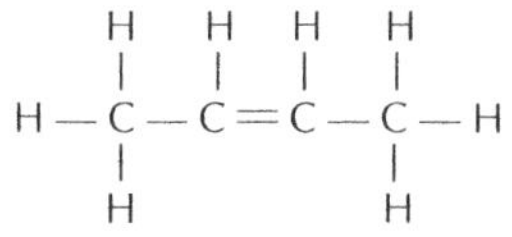	But-2-ene	
Alkynes	$CH{\equiv}CCH_2CH_3$ 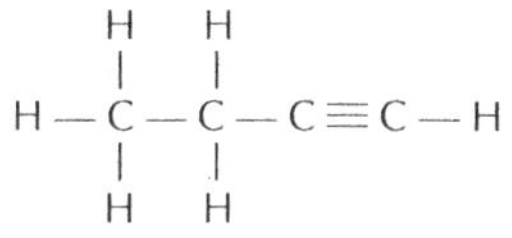	But-1-yne	
	$CH_3C{\equiv}CCH_3$ 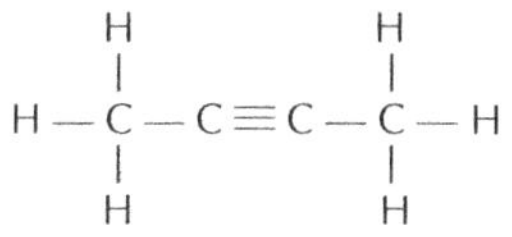	But-2-yne	
Alcohols	$CH_3CH_2CH_2OH$ 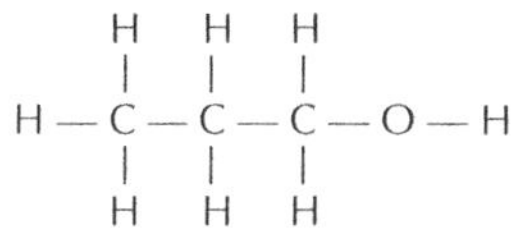	Propan-1-ol	Note that we also add in an 'an' in the name as we could get an alcohol with a double bond in it making it propen-1-ol!
	$CH_3CH_2CH(OH)CH_3$ 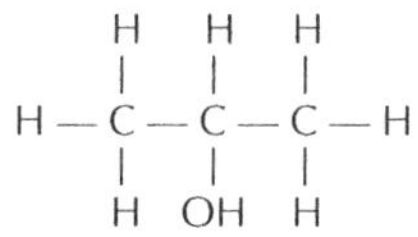	Propan-2-ol	We write things that are branched off from the main chain other than hydrogens in brackets.
Haloalkanes	$CH_3CH_2CH_2Cl$ 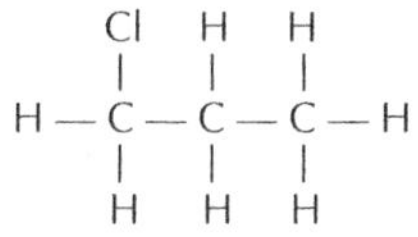	1-chloropropane	The number goes in front of the functional group always even if the functional group is in the beginning of the name.
Amines	$CH_3CH_2CH_2NH_2$ 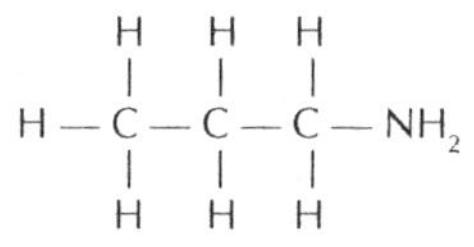	1-aminopropane or propan-1-amine	

We never number carboxylic acids as they are always on carbon number 1.

What do the following key words or phrases mean?

Organic chemistry	
Alkane	
Alkene	
Alkyne	
Haloalkane	
Alcohol	
Amine	
Carboxylic acid	
Functional group	

2.4: Numbering and naming

<table>
<tr><th></th><th>Structure</th><th>Name</th></tr>
<tr><td>1</td><td><pre>
    H   H   H   H
    | | | |
H — C — C — C — C — O — H
    | | | |
    H   H   H   H
</pre></td><td></td></tr>
<tr><td>2</td><td><pre>
    H   H   H   H   H
    | | | | /
H — C — C — C — C — N
    | | | | \
    H   H   H   H   H
</pre></td><td></td></tr>
<tr><td>3</td><td><pre>
    H   H   H   H
    | | | |
H — C — C — C — C = C
    | | | | |
    H   H   H   H   H
</pre></td><td></td></tr>
<tr><td>4</td><td><pre>
    H   H   H   H
    | | | |
H — C — C — C = C — C — H
    | | | |
    H   H   H   H
</pre></td><td></td></tr>
<tr><td>5</td><td>$CHCCH_2CH_2CH_3$</td><td></td></tr>
<tr><td>6</td><td>$CH_3CH_2CH_2CH_2OH$</td><td></td></tr>
<tr><td>7</td><td>$CH_3CH_2CH_2CH_2CH_2CH_2CH_2Cl$</td><td></td></tr>
<tr><td>8</td><td>$CH_3CH(OH)CH_2CH_3$</td><td></td></tr>
</table>

 ISBN: 9780170260107

If we have branches coming off the main chain which are not a functional group, these must be numbered also. This number tells us what carbon number the chain comes off in relation to the functional group.

Example 1: 1 branch off an alkane

```
      H   H   H   H
      |   |   |   |
  H — C — C — C — C — H
      |   |   |   |
      H   |   H   H
      H — C — H
          |
          H
```

Step 1: There are four carbons in the longest chain so we know that the prefix is 'but'.
Step 2: The functional group is an alkane since there are only carbons and hydrogens attached; so the suffix is -ane.
Step 3: We don't need to number an alkane because it is not a functional group.
Step 4: We call a CH_3 branch methyl since there is one carbon in it and any alkane branch chain has the ending –yl. We number this branch at the beginning of the name.
This gives us: 2-methylbutane

It is important to not put in any spaces between the words. You must put a hyphen in between the number and the branch. For example: 2-methylbutane.

Example 2: 1 branch off another functional group

```
      H   H   H   H
      |   |   |   |
  H — C — C — C — C — O — H
      |   |   |   |
      H   |   H   H
      H — C — H
          |
          H
```

Step 1: There are four carbons in the longest chain so we know that the prefix is 'but' again.
Step 2: The functional group is an alcohol since there is an OH; so the suffix is -ol.
Step 3: The functional group gets the lowest possible number; and so we always number it first, hence butan-1-ol.
Step 4: The branch then is still a methyl group but as the carbon attached to the OH is carbon 1, the methyl group is attached to carbon number 3.
This gives us: 3-methylbutan-1-ol

Example 3: 2 branches

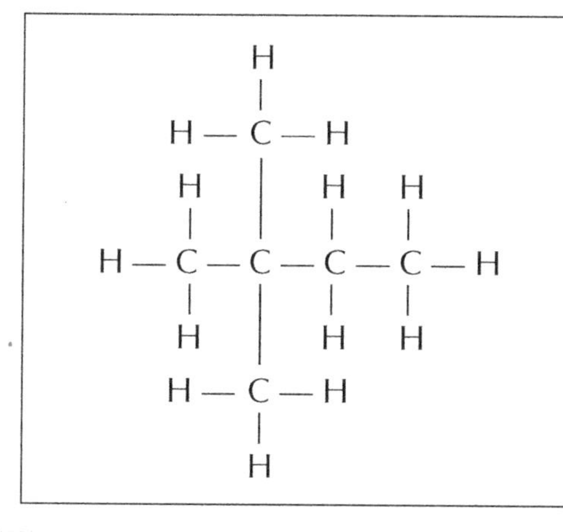

Step 1: But–
Step 2: -ane
Step 3: No number required.
Step 4: There are two methyl groups attached this time and we have to number them both separately even though they are on the same carbon. We also add the prefix di- to methyl to tell us we have two of them, (if we had three we would put tri- in front).
This gives us: 2,2-dimethylbutane

Example 4: What is the longest chain?

```
CH3 — CH2
        |
      CH2 — CH3
        |
      CH3
```

Step 1: Despite the bend we still have four carbons in the longest chain so we still have the prefix 'but'.
Step 2: -ane
Step 3: No number required
Step 4: Methyl
This gives us: 2-methylbutane

ISBN: 9780170260107

Example 5: What comes first?

$CH_3CH(CH_3)CH_2Cl$

Step 1: Prop
Step 2: Haloalkane so we have chloro- in the beginning and -ane as the suffix.
Step 3: 1-chloro, as it gets priority with its numbering since it is the functional group.
Step 4: Methyl–, we put this after the 1-chloro since 'c' comes first in the alphabet. **This gives us:** 1-chloro-2-methylpropane

2.5: Naming branched molecules

Name the following branched molecules.

	Organic structure	Name
1	Structural formula: H₂C=CH–CH₃ drawn with all H atoms ($CH_2{=}CH{-}CH_3$)	
2	$CH_3CH(CH_3)CH(CH_3)CH_3$ (drawn as $CH_3CHCHCH_3$ with CH_3 above and below)	
3	$CH_3CH(OH)CH(CH_3)CH_2CH_3$	
4	$CH_3CH(Cl)CH(CH_3)CH_2CH_3$	
5		2,3,3-trimethylhexane
6		2-methylpropan-2-ol
7		2,2,3,3-tetramethylbutane

 ISBN: 9780170260107

INVESTIGATION 1

Molecules in three dimensions

AIM: To build three-dimensional models of and name organic molecules.

EQUIPMENT
Molymods or plasticine and sticks

METHOD
Make the following models in three dimensions making sure the atoms are spaced as far apart as possible. Fill in the rest of the table.

	Name	Structural formula	Condensed structural formula
1	Methane	H | H — C — H | H	CH_4
2	Ethane		
3	Ethene		
4	Ethyne		

ISBN: 9780170260107

	Name	Structural formula	Condensed structural formula
5	Ethanol		
6	Chloroethane		
7	Ethanamine or aminoethane		
8	Methanoic acid		
9	2-methylbutane		

QUESTIONS

1 For carbon atoms with single bonds, what is the angle between these bonds?

2 What does this tell you about the shape of the molecule?

ISBN: 9780170260107

Empirical and molecular formula

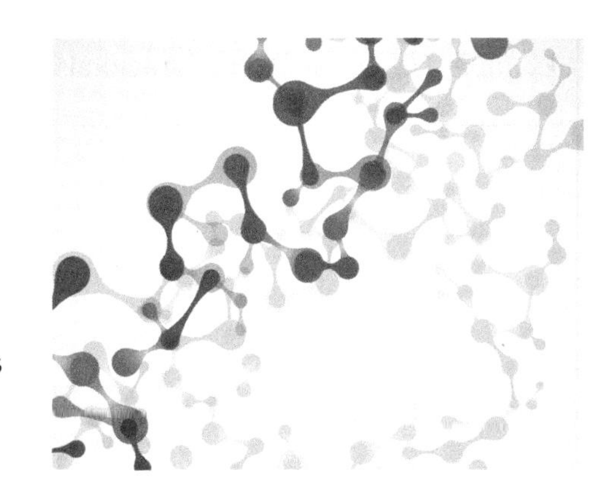

The molecular formula of a compound tells you how many of each atom is present in that molecule but does not tell you how the atoms are arranged, for example C_6H_{12}. You can tell from this formula that it is an alkene since there are not enough hydrogens for every carbon atom to be fully surrounded by them; but you can't tell anything else, such as whether that molecule has branches coming off it.

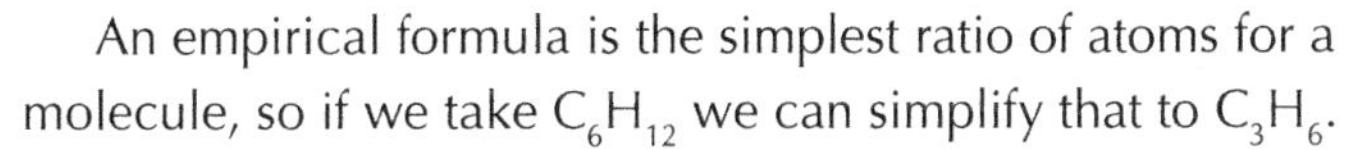

An empirical formula is the simplest ratio of atoms for a molecule, so if we take C_6H_{12} we can simplify that to C_3H_6.

2.6: Molecular and empirical formula

Complete the table below, writing down the molecular formula and empirical formula for the following organic substances. The first two have been done to help you.

	Condensed structural formulas	Molecular formula	Empirical formula
1	$CH_3CH_2CH_3$	C_3H_8	C_3H_8
2	$CH_3CH_2CH_2CH_3$	C_4H_{10}	C_2H_5
3	$CHCCH_2CH_2CH_3$		
4	$CH_3CH_2CH_2CH_2OH$		
5	$CH_3CH_2CH_2CH_2CH_2CH_2CH_2Cl$		
6	$CH_3CH(OH)CH_2CH_3$		
7	$CH_3CH(OH)CH(CH_3)CH_2CH_3$		
8	$CH_3CH(Cl)CH(CH_3)CH_2CH_3$		

Structural isomers

These are molecules with the same molecular formula but a different arrangement in space, just like from one piece of paper you can get many different shapes using Origami – the Japanese art of paper folding.

FACT

The first person to realise that not all organic chemicals came from living things was Friedrich Wöhler. Before him it was believed that compounds that come from living things were distinctive. He manufactured with a few household chemicals urea (a waste product from animals found in urine). What he didn't know at the time was that he actually produced an isomer of urea – he made NH_4CNO instead of $CO(NH_2)_2$. He discovered this later by looking at the properties of the two chemicals. This was the beginning of organic chemistry.

ISBN: 9780170260107

Example 1: What are all the structural isomers of C_4H_{10}?

```
    H   H   H   H
    |   |   |   |
H — C — C — C — C — H
    |   |   |   |
    H   H   H   H
```

There are two structural isomers of C_4H_{10}: the first is butane which is shown to the left. The second is 2-methylpropane which is shown to the right.

```
    H   H   H
    |   |   |
H — C — C — C — H
    |   |   |
    H   |   H
        |
    H — C — H
        |
        H
```

There are only two because if we add a methyl group to the end of propane we get butane again (remember the longest chain can bend).

Example 2: What are all the structural isomers of C_5H_{12}?

There are three isomers this time:

```
    H   H   H   H   H
    |   |   |   |   |
H — C — C — C — C — C — H
    |   |   |   |   |
    H   H   H   H   H
```

Pentane

```
    H   H   H   H
    |   |   |   |
H — C — C — C — C — H
    |   |   |   |
    H   H   |   H
            |
        H — C — H
            |
            H
```

2-methylbutane

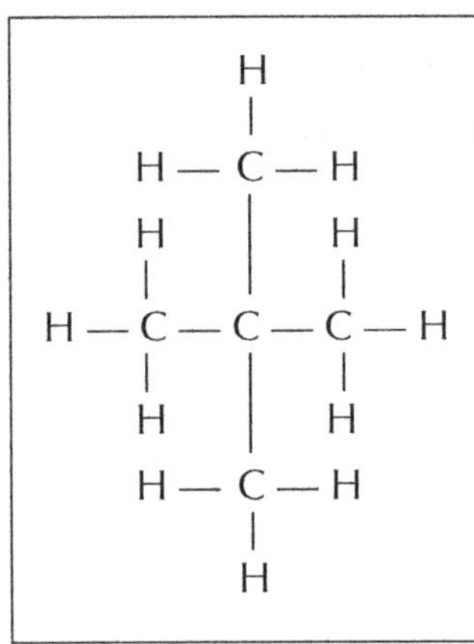

2,2-dimethylpropane

2.7: Structural isomers

1 Draw all the structural isomers of C_6H_{14}.

EXAMINER'S TIP

Take care! Count all the hydrogens and carbons to make sure your isomer adds up to the molecular formula.

 ISBN: 9780170260107

2 Draw all the structural isomers of C_7H_{14}.

Geometric isomers

These are a special type of structural isomer which only occur in alkenes. They exist because there is no free rotation about a double bond, which means the bonds attached to the carbons in a double bond can't move about. They are fixed in position, giving us two sorts of isomers, **cis** and **trans**.

Example:

CH_3 \ C=C / CH_3; H / C=C \ H	CH_3 \ C=C / H; H / C=C \ CH_3
*Cis*but-2-ene	*Trans*but-2-ene

As there are two different groups attached to each carbon in the double bond they can either be arranged in a cis arrangement, or on the same side, or trans arrangement or on different sides.

Geometric isomers form for these two reasons:

1 There is no free rotation about a double bond.

2 There are two different groups attached to each carbon in the double bond.

2.8: Geometric isomers

Decide whether the following alkenes will form geometric isomers and if they do, name them and draw out their structural formulas.

	Condensed structural formula	Does this molecule form geometric isomers?	Structural formulas and names of the isomers if they form
1	HC(Cl)=C(Cl)H	Yes	H \ C=C / Cl; Cl / C=C \ H — *trans*-1,2-dichloroethene H \ C=C / H; Cl / C=C \ Cl — *cis*-1,2-dichloroethene

ISBN: 9780170260107

2	$CH_3CH_2CH{=}CH_2$		
3	$CH_3C(Cl){=}C(Cl)CH_3$		
4	HC(Br)=C(Br)H		

What do the following key words mean?

Structural isomer	
Geometric isomer	
Molecular formula	
Empirical formula	
Structural formula	

CHECKPOINT 1
WHAT HAVE YOU LEARNED SO FAR?

Naming and isomerism

1 Circle and name all the functional groups you find in the following molecules.

a Aspirin

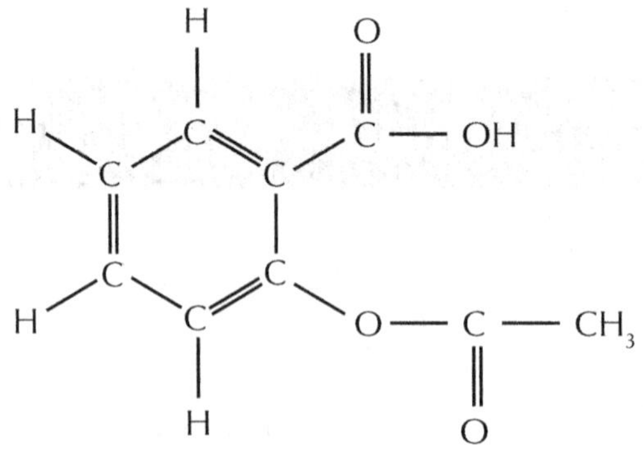

Note: There are extra groups here that you don't need to know about this year.

ISBN: 9780170260107

b Viagra

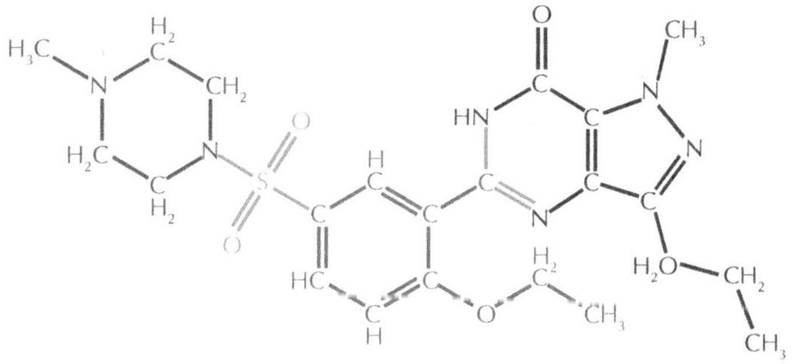

2 Fill in the table below.

	Name	Condensed structural formula	Molecular and empirical formula	Functional group
a	2,2-dimethylbutane			Alkane
b		$CH_3CH(Cl)CH_3$	C_3H_7Cl	
c		CH_3COOH		
d	2-methylpropan-1-ol			

3 Explain why but-1-ene can't form geometric isomers but is a structural isomer to but-2-ene which can form geometric isomers.

Alkanes

The alkanes are a family of hydrocarbons with hydrogen and carbon atoms all **singly** bonded. They are well-known for their use as fuels as they combust well, for example octane is the main component in petrol. They undergo very few reactions as they are quite stable and don't have a reactive part or functional group. Alkanes are considered **saturated molecules** as they have the maximum number of bonds attached to each carbon, with no double or triple bonds.

Alkanes have **low** melting points and boiling points as they form **weak intermolecular bonds** between their molecules. The longer the chain the greater the number of intermolecular bonds and so the higher the melting point and boiling point. They will only dissolve in **non-polar solvents** as they are non-polar molecules.

Reactions of alkanes

The main reaction alkanes undergo are called **substitution reactions** as one hydrogen atom is exchanged for another atom. These reactions are *very slow* and often require a *catalyst.*

Substitution to form haloalkanes

A **haloalkane** is an organic compound with a halogen (F, Cl, Br or I) attached to it.

To form a haloalkane from an alkane, ***UV light*** is required and the reaction takes place slowly. In light, bromine water (Br_2), an **orange solution**, changes and becomes **colourless** because it reacts with methane. Chlorine water is colourless, however in gaseous form chlorine is pale green.

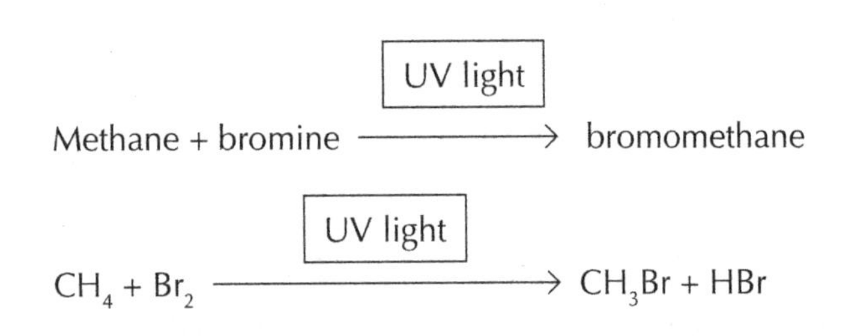

Methane + bromine $\xrightarrow{\text{UV light}}$ bromomethane

$CH_4 + Br_2 \xrightarrow{\text{UV light}} CH_3Br + HBr$

When you define what a substitution reaction is in the exam make sure you do NOT use the word *substitute* to define it, use the word **exchange** instead.

2.9

Substitution reactions of alkanes

Complete the following reactions to show the products or reactants. The first one has been done for you.

1 $CH_3CH_3 + Cl_2 \xrightarrow{\text{UV light}} CH_3CH_2Cl$

2 $CH_3CH_3 + Br_2 \xrightarrow{\text{UV light}}$ ____________________

3 $CH_3CH_2CH_2CH_3 + Cl_2 \xrightarrow{\text{UV light}}$ ____________________

4 $CH_3CH_2CH_2CH_2CH_3 + Br_2 \xrightarrow{\text{UV light}}$ ____________________

5 $CH_3CH_2CH_2CH_2CH_2CH_2CH_3 + Cl_2 \xrightarrow{\text{UV light}}$ ____________________

ISBN: 9780170260107

Alkenes

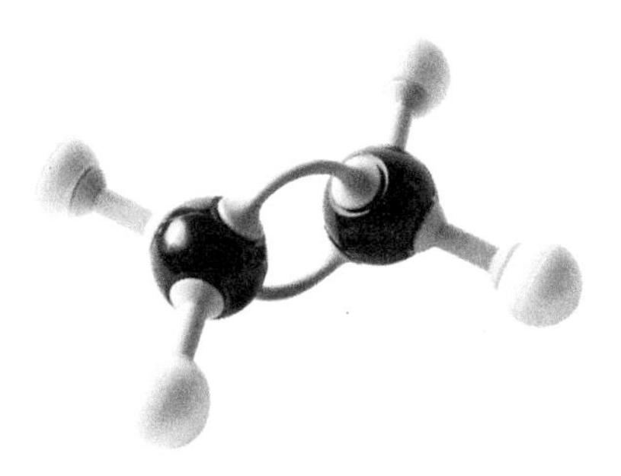

The alkene family also has only hydrogen and carbon atoms, however they contain at least one **double bond** between carbon atoms. This is the reactive part of the molecule and because of this, alkenes have a functional group, unlike alkanes. Alkenes are considered unsaturated since they contain a double bond, meaning they don't have the maximum number of bonds possible for an organic molecule.

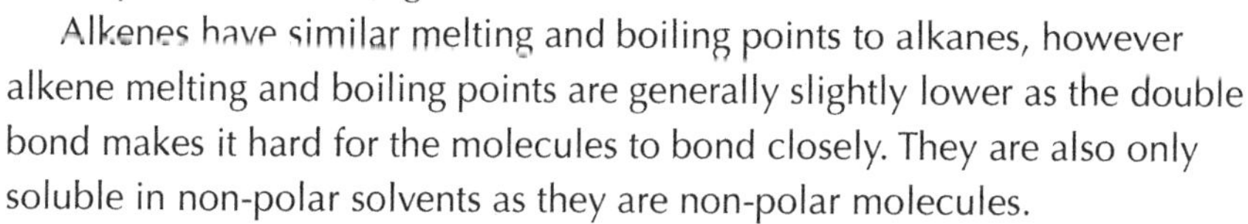

Alkenes have similar melting and boiling points to alkanes, however alkene melting and boiling points are generally slightly lower as the double bond makes it hard for the molecules to bond closely. They are also only soluble in non-polar solvents as they are non-polar molecules.

Reactions of alkenes

Oxidation using H^+/MnO_4^-, acidified permanganate

We can oxidise alkenes using permanganate to make a diol. This means an OH group is added to each carbon atom in the double bond, and the double bond is lost. Like all organic reactions you only need to write the organic product so don't worry about the product of the acidified permanganate. The visible change in this reaction is purple to colourless as the acidified permanganate is purple and the product it forms, the manganese ion, Mn^{2+}, is colourless.

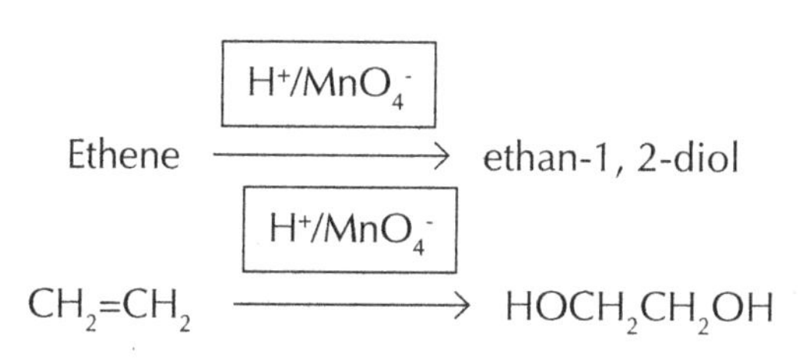

Addition

Alkenes undergo many addition reactions. In addition reactions the double bond is lost and an atom or group of atoms is added on to either side of the double bond. These reactions usually have a fast reaction rate as the double bond is quite reactive.

With H_2 and a platinum catalyst

These addition reactions form alkanes.

$$CH_2{=}CH_2 + H_2 \xrightarrow{\boxed{Pt}} CH_3CH_3$$

With Cl_2 or Br_2

These addition reactions form haloalkanes.

$$CH_2CH_2 + Br_2 \longrightarrow BrCH_2CH_2Br$$

The visible change during this reaction is orange to colourless as bromine water is orange in colour, and none of the products formed have a colour.

$$CH_2CH_2 + Cl_2 \longrightarrow ClCH_2CH_2Cl$$

In this reaction colour change will only occur if you are using gaseous chlorine which is pale green, otherwise chlorine water is colourless and all the products are colourless.

With H_2O and dilute sulfuric acid catalyst

These addition reactions form alcohols.

$$CH_2CH_2 + H_2O \xrightarrow{\boxed{H^+}} CH_3CH_2OH$$

With HX

These addition reactions form haloalkanes, with 'X' being Cl or Br.

$$CH_2CH_2 + HCl \longrightarrow CH_3CH_2Cl$$

Major and minor products

If an alkene molecule is symmetrical, like ethane, there is only one product with water, as shown in the alcohol-forming reaction above. However if an alkene is asymmetrical, like propene, the OH or X could add on to either carbon in the double bond and form a different molecule.

$$CH_3CH{=}CH_2 + H_2O \xrightarrow{\boxed{H^+}} CH_3CH(OH)CH_3 + CH_3CH_2CH_2OH$$

$$\text{Propene + water} \xrightarrow{\boxed{H^+}} \text{Propan-2-ol + propan-1-ol}$$

$$CH_3CH{=}CH_2 + HBr \longrightarrow CH_3CH(Br)CH_3 + CH_3CH_2CH_2Br$$

$$\text{Propene + hydrogen bromide} \longrightarrow \text{2-bromopropane + 1-bromopropane}$$

Propan-2-ol and 2-bromopropane turn out to be the major products as organic molecules will generally add the extra hydrogen onto the atom with the most hydrogen atoms (in this case the CH_2). This rule, named after Markovnikov, states that the ***'rich get richer'***, in other words a hydrogen-rich carbon atom in a double bond will usually get richer over a hydrogen-poor carbon atom. Therefore we say the propan-1-ol is the minor product as less of this product will form.

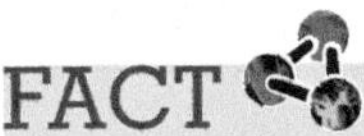

Vladimir Vasilyevich Markovnikov was a Russian chemist who came up with the addition to alkenes rules in 1869. He originally studied economics and later changed to chemistry.

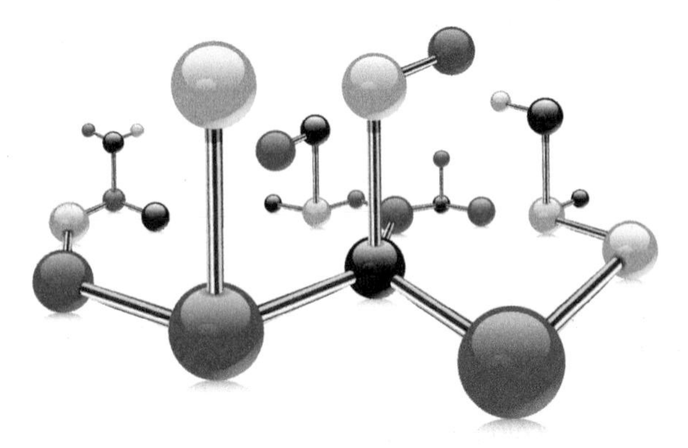

Polymerisation

A special kind of addition reaction is called **polymerisation**. A polymer is a large organic molecule made up of repeating units called monomers. With alkenes the double bond is broken down to form two new single bonds to attach to many more of the same alkene. Conditions vary depending on the alkene needed to be made into a polymer, but usually a catalyst, high temperature and pressure are used.

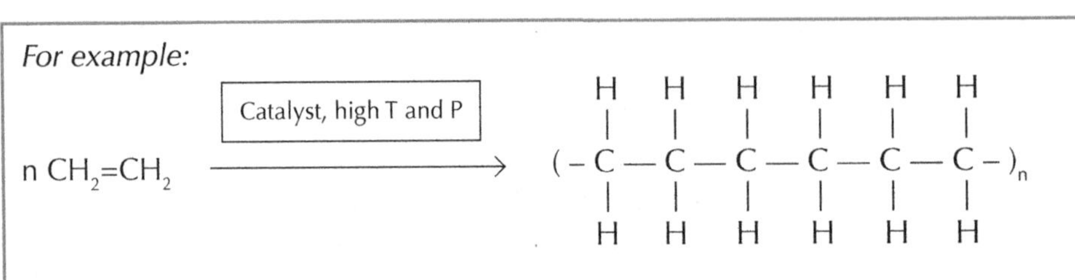

Note: 'n' means many alkenes; depending on the product made this number can vary widely.

ISBN: 9780170260107

2.10: Reactions of alkenes

Write down the products made from the following reactions of alkenes. The first two have been done for you.

1 $CH_3CH{=}CH_2 \xrightarrow{H^+/MnO_4^-} CH_3CH_2(OH)CH_2OH$

2 $CH_3CH{=}CH_2 + Br_2 \longrightarrow CH_3CH(Br)\,CH_2Br$

3 $CH_3CH{=}CH_2 + Cl_2 \longrightarrow$ ....

4 $CH_3CH{=}CH_2 + H_2O \xrightarrow{H^+}$ ....

5 $nCH_3CH{=}CH_2 \xrightarrow{\text{Catalyst, high T and P}}$ ....

6 $CH_3CH{=}CH_2 + H_2 \xrightarrow{Pt}$ ....

7 $CH_3CH{=}CHCH_3 + HBr \longrightarrow$ ....

8 $CH_3CH{=}CHCH_3 + H_2O \xrightarrow{H^+}$ ....

9 $CH_3CH{=}CHCH_3 \xrightarrow{H^+/MnO_4^-}$ ....

10 $n\ CH_3CH{=}CHCH_3 \xrightarrow{\text{Catalyst, high T and P}}$ ....

CHEMISTRY APPS

Your fruit is ripened on the way to the supermarket by ethene gas, otherwise known as ethylene. This gas is naturally released by fruit over time, which is why you shouldn't keep your bananas in your fruit bowl if you want your other fruit to last a long time. As bananas mature they release this chemical, which turns them to yellow (and then brown) and gives them that nice taste. However it will also mature your other fruit.

ISBN: 9780170260107

What do the following key words mean?

Substitution reaction	
Oxidation reaction	
Addition reaction	
Polymerisation reaction	
Markovnikov's rule	
Saturated molecule	
Unsaturated molecule	

EXPERIMENT 1

Properties and reactions of alkanes and alkenes

AIM: To compare the properties and reactions of alkanes and alkenes.

EQUIPMENT and CHEMICALS:

cyclohexane
cyclohexene
0.02 mol L^{-1} acidified potassium permanganate, $H^+/KMnO_4$
bromine water, Br_2
test tubes
test tube rack

SAFETY PRECAUTIONS:

Both cyclohexane and cyclohexene are slight irritants to skin and are flammable. Potassium permanganate is an irritant to eyes and skin but is not flammable. Bromine water is corrosive to the skin and is also a skin and eye irritant.

METHOD and RESULTS:

	Observations		
Hydrocarbon	Add 1 mL of acidified potassium permanganate	Add 1 mL of bromine water	Solubility in water (add a few mLs of water and see if the hydrocarbon dissolves)
To 1 mL of cyclohexane			
To 1 mL of cyclohexene			

ISBN: 9780170260107

QUESTIONS:

1 Complete the reaction below to show the reaction between cyclohexene and acidified potassium permanganate.

$H^+MnO_4^-$ →

2 What kind of reaction is this? ____________________

____

3 Complete the reaction below to show the reaction between cyclohexane and bromine water.

Br_2/UVlight →

4 What kind of reaction is this? ____________________

____

5 Complete the reaction below to show the reaction between cyclohexene and bromine water.

Br_2 →

6 What kind of reaction is this? ____________________

____

7 Why are neither of the above hydrocarbons soluble in water?

____

____

CHECKPOINT 2

WHAT HAVE YOU LEARNED SO FAR?

Hydrocarbons

1 Fill in the reaction scheme below of ethane reacting with bromine water.

Name: Ethane

Structural formula:

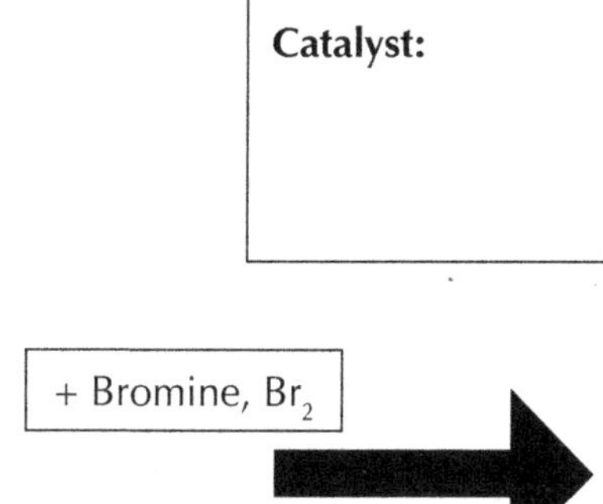

Name of product:

Structural formula:

ISBN: 9780170260107

What observations would you make as this reaction proceeds?

2 Fill in the reaction scheme below of all the reactions propene undergoes.

Type of reaction: Oxidation

Name of product: Propan-1,2-diol

Structural formula:

Type of reaction: Addition

Name of product:

Structural formula:

Type of reaction: Addition

Name of product:

Structural formula:

```
    H   H   H
    |   |   |
H — C — C — C — H
    |   |   |
    H   H   H
```

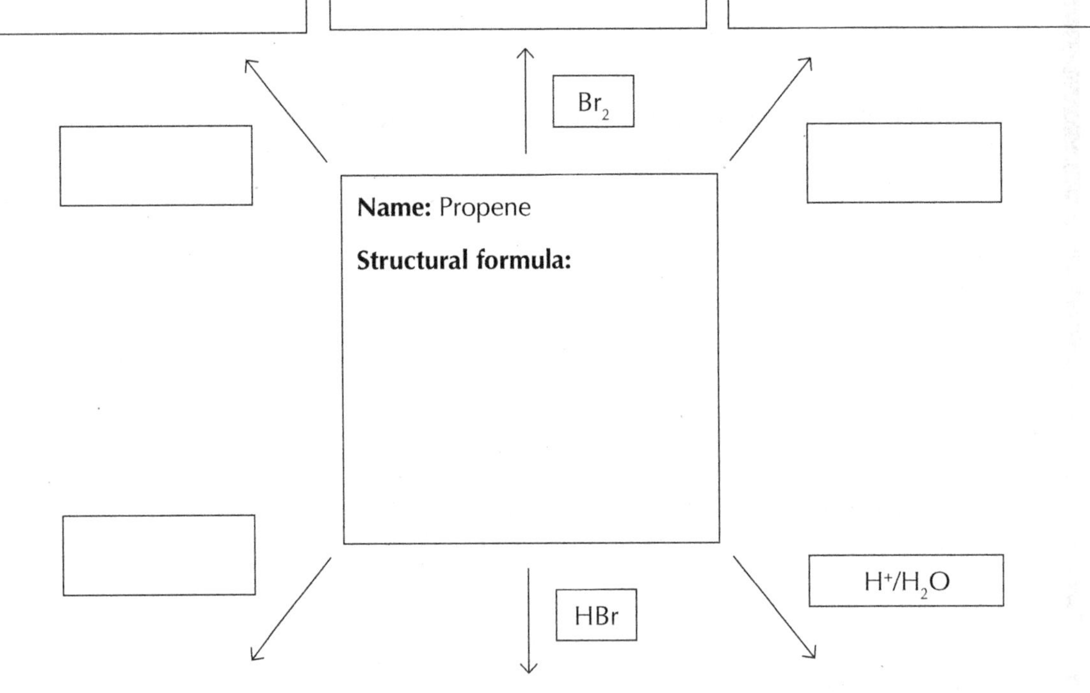

Type of reaction: Polymerisation

Name of product: Polypropylene

Structural formula:

Type of reaction:

Name of product:

Structural formula:

Type of reaction:

Name of product: Propan-1-ol and Propan-2-ol

Structural formula:

ISBN: 9780170260107

Alcohols

Alcohols have a higher boiling point and melting point than hydrocarbon molecules of similar size, as alcohols have a polar OH group(s). This makes the shorter chained alcohols soluble in water and polar solvents, however as the chains get longer they are much less soluble in water.

Classification

We can classify alcohols as primary, secondary or tertiary according to where the functional group is placed.

A **primary alcohol** is one which has the alcohol group attached to a carbon that has one other carbon attached, like propan-1-ol shown below.

```
      H   H   H
      |   |   |
  H — C — C — C — O — H
      |   |   |
      H   H   H
```

A **secondary alcohol** is one which has the alcohol group attached to a carbon that is attached to two other carbon atoms, like in propan-2-ol shown below.

```
      H   H   H
      |   |   |
  H — C — C — C — H
      |   |   |
      H   OH  H
```

A **tertiary alcohol** is one which has the alcohol group attached to a carbon that is attached to three other carbon atoms, like methylpropan-2-ol shown below.

```
      H  CH3  H
      |   |   |
  H — C — C — C — H
      |   |   |
      H   OH  H
```

ISBN: 9780170260107

Reactions of alcohols

Substitution to form haloalkanes

Three reagents can be used to substitute an alcohol functional group with a chlorine atom. They are PCl_3, PCl_5 and $SOCl_2$.

```
    H   H   H                                          H   H   H
    |   |   |         PCl3 or PCl5 or SOCl2            |   |   |
H — C — C — C — O — H  ————————————————————→       H — C — C — C — Cl
    |   |   |                                          |   |   |
    H   H   H                                          H   H   H

   Propan-1-ol                                      1-chloropropane
```

Oxidation to form carboxylic acids

Either acidified permanganate (H^+/MnO_4^-) or acidified dichromate ($H^+/Cr_2O_7^{2-}$) can be used to turn a primary alcohol into a carboxylic acid. The acidified permanganate will change from purple to colourless and the acidified dichromate will change from an orange solution to a green one. You will need to remember these observations.

```
    H   H   H                                       H   H   H
    |   |   |        H+/MnO4- or H+/Cr2O7 2-        |   |   |   //O
H — C — C — C — O — H  ——————————————————→      H — C — C — C — C
    |   |   |                                       |   |   |   \
    H   H   H                                       H   H   H    O — H

   Propan-1-ol                                       Propanoic acid
```

Elimination to form alkenes

Using concentrated sulfuric acid we can remove water from an alcohol to form an alkene.

```
    H   H   H                                       H   H   H
    |   |   |             Conc. H2SO4               |   |   |
H — C — C — C — O — H  ——————————————————→      H — C — C = C
    |   |   |                                       |   |   |
    H   H   H                                       H   H   H

   Propan-1-ol                                        Propene
```

Only one product forms when you have a primary alcohol, however when you have a secondary or tertiary alcohol two products can form. For example, there are two products shown at the top of the next page when pentan-2-ol is dehydrated. The first product pent-1-ene is the minor product as the carbon atom on the end is unlikely to lose a hydrogen compared to a carbon in the middle of the chain, so pent-2-ene will be the major product (the most common product).

ISBN: 9780170260107

```
                  H
                  |
   H   H   H   O   H        Conc. H2SO4        H   H   H       H            H           H   H
   |   |   |   |   |                           |   |   |       |            |           |   |
H— C — C — C — C — C —H    ————————————>    H— C — C — C — C = C       H— C — C = C — C — C —H
   |   |   |   |   |                           |   |   |   |   |            |   |   |   |   |
   H   H   H   H   H                           H   H   H   H   H +          H   H   H   H   H
```

Pentan-2-ol

Minor Product: Pent-1-ene + **Major Product:** Pent-2-ene

2.11: Classification and reactions of alcohols

1 Classify the following alcohols as primary, secondary or tertiary. The first one has been completed for you.

Alcohol	$CH_3-CH(OH)-CH_2-CH_3$	$H-CH_2-CH_2-O-H$ (H–C(H)(H)–C(H)(H)–O–H)	$H-C(H)(H)-C(H)(H)-C(H)(H)-C(H)(H)-OH$
Classification	Secondary		

Alcohol	$CH_3-CH(CH_3)-CH_2-CH_2-OH$	$CH_3-CH_2-C(CH_3)(OH)-CH_2-CH_3$	$H-C(H)(H)-C(H)(H)-C(H)(OH)-C(H)(H)-C(H)(H)-H$
Classification			

2 Complete the reactions these alcohols undergo.

a

```
   H   H
   |   |          H+/MnO4-
H— C — C — O — H  ————————>
   |   |
   H   H
```

b

```
   H   H
   |   |          PCl3
H— C — C — O — H  ————————>
   |   |
   H   H
```

c $CH_3-CH(OH)-CH_2-CH_3$ $\xrightarrow{\text{Conc. } H_2SO_4}$

d $CH_3-CH(OH)-CH_2-CH_3$ $\xrightarrow{H^+/Cr_2O_7^{2-}}$

e

```
    H   H  OH  H   H
    |   |   |  |   |
H — C — C — C — C — C — H
    |   |   |  |   |
    H   H   H  H   H
```

$\xrightarrow{SOCl_2}$

f

```
    H   H  OH  H   H
    |   |   |  |   |
H — C — C — C — C — C — H
    |   |   |  |   |
    H   H   H  H   H
```

$\xrightarrow{\text{Conc. } H_2SO_4}$

CHEMISTRY APPS

Methanol or CH_3OH is a common product when making 'moonshine' liquor (alcohol made at home). It is deadly in small quantities. To treat for methanol poisoning, in a hospital, ethanol (the standard alcohol in alcoholic drinks) is given, as your body preferentially processes it over the methanol, allowing the methanol to exit the body without being broken down.

EXPERIMENT 2

Solubility and oxidation of alcohols

AIM: To investigate the reaction of an alcohol with oxidants and to investigate the solubility of alcohols.

EQUIPMENT and CHEMICALS:

test tubes
ethanol
0.1 mol L^{-1} acidified potassium dichromate, $H^+/K_2Cr_2O_7$
0.02 mol L^{-1} acidified potassium permanganate, $H^+/KMnO_4$
250 mL beaker

SAFETY PRECAUTIONS:

Potassium permanganate is an irritant to skin and eyes. Potassium dichromate is a very hazardous skin and eye irritant. Ethanol is flammable.

METHOD and RESULTS:

Alcohol	Observations		
	1 mL of $H^+/Cr_2O_7^{2-}$ added	1 mL of H^+/MnO_4^- added	Solubility in a few mLs of water
To 1 mL of ethanol			

 ISBN: 9780170260107

QUESTIONS:

1 Complete the following reactions of ethanol.

a $CH_3CH_2OH \xrightarrow{H^+/Cr_2O_7^{2-}}$ ....

b $CH_3CH_2OH \xrightarrow{H^+/MnO_4^-}$ ....

2 Why does ethanol dissolve in water?

....

....

....

....

....

....

....

Haloalkanes

Haloalkanes have boiling and melting points in between alcohols and alkanes, as the halogen attached makes a polar bond with carbon. There is a smaller difference in electronegativity than between the oxygen and hydrogen atoms in alcohols. They also can be classified as primary, secondary and tertiary in the same way that alcohols can.

Haloalkanes are used in aerosols like this.

Reactions of haloalkanes

Substitution to form alcohols

Using aqueous potassium hydroxide the halogen can be exchanged with an OH to form an alcohol.

$$CH_3CH_2Br \xrightarrow{KOH\ (aq)} CH_3CH_2OH$$

(Structural formulas shown: H–C(H)(H)–C(H)(H)–Br → H–C(H)(H)–C(H)(H)–O–H)

Substitution to form amines

Using alcoholic or aqueous ammonia (that's ammonia dissolved in alcohol) the halogen can be exchanged with an NH_2 to form an amine.

$$CH_3CH_2Br \xrightarrow{NH_3\ (alc)} CH_3CH_2\text{–}NH_2$$

(Structural formula shown: H–C(H)(H)–C(H)(H)–Br)

Elimination to form alkenes

Using concentrated sodium hydroxide or potassium hydroxide in alcohol and heat (in reflux conditions or gentle heating) we can remove the halogen and a corresponding hydrogen atom to form an alkene. Major and minor products can form here too if the haloalkane is asymmetric.

```
    H   H                                        H       H
    |   |                                         \     /
H — C — C — Br   Conc. NaOH (alc) and reflux        C=C
    |   |        ─────────────────────────→       /     \
    H   H                                        H       H
```

CHEMISTRY APPS

Chlorofluorocarbons, better known as CFCs, are small organic molecules with chlorine and fluorine atoms. CFCs are highly useful in aerosols, foam plastics, air-conditioners, fridges and fire extinguishers – but in the 1970s these gases were discovered to be destroying the ozone layer (ozonosphere) that protects the world from UV light. CFCs are now banned in most countries.

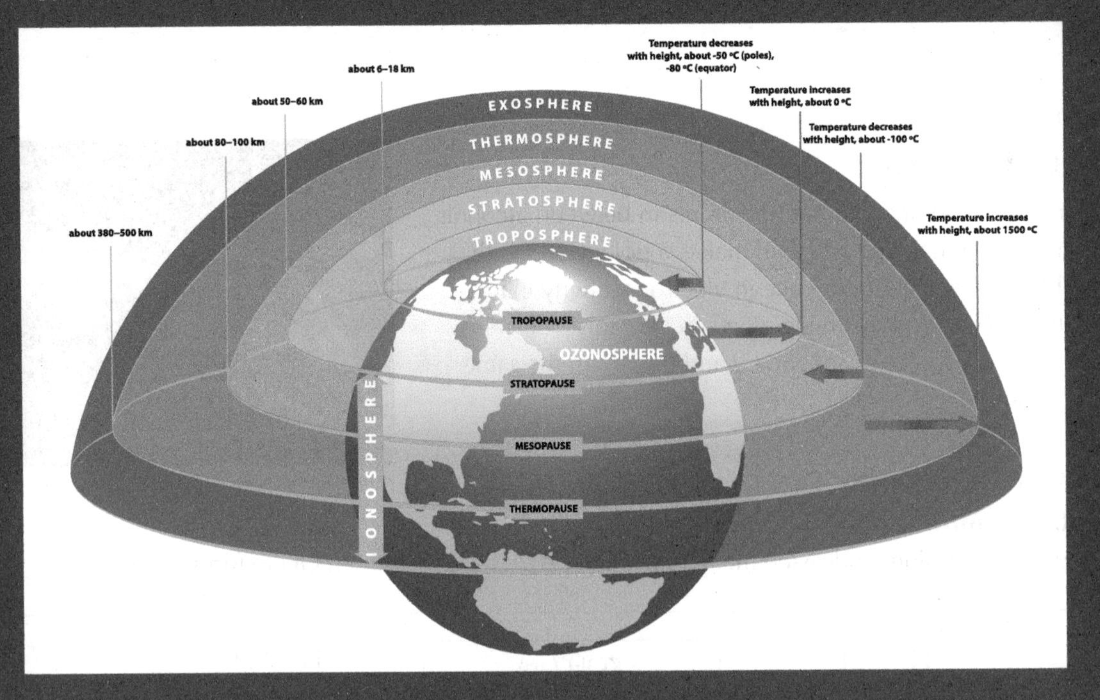

2.12: Classification and reactions of haloalkanes

1 Classify the following haloalkanes as primary, secondary or tertiary. The first one has been completed for you.

Haloalkane	Br on C-3 of $CH_3CH_2CHCH_3$ (Br above the CH)	H H H H—C—C—C—Cl H H H	H Cl H H—C—C—C—H H H H
Classification	Secondary		

 PHOTOCOPYING OF THIS PAGE IS RESTRICTED UNDER LAW. ISBN: 9780170260107

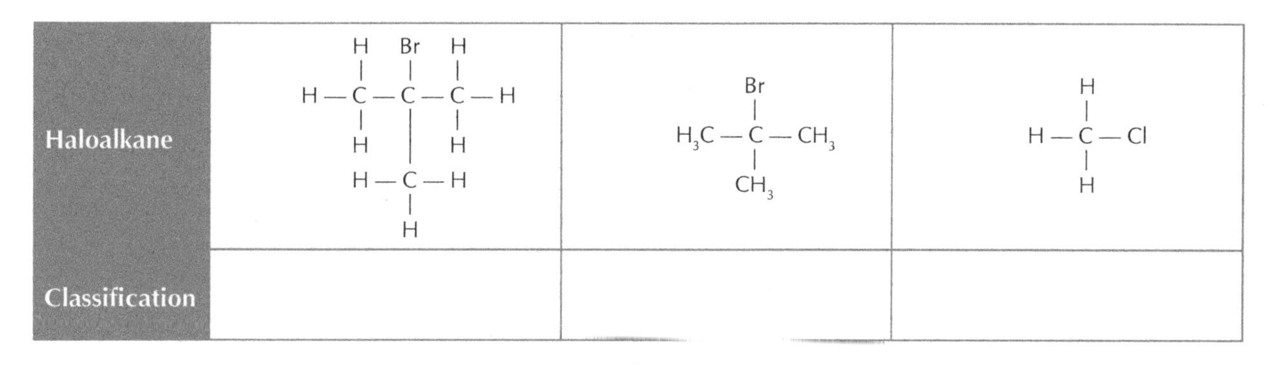

Haloalkane	H Br H H — C — C — C — H H │ H H — C — H H	Br H_3C — C — CH_3 CH_3	H H — C — Cl H
Classification			

2 Complete the reactions these haloalkanes undergo below.

a $CH_3CH_2CHBrCH_3$ (Br on the third carbon) —[Conc. NaOH (alc) and reflux]→ ....

b $CH_3CH_2CHBrCH_3$ (Br on the third carbon) —[NH_3 (alc)]→ ....

c H — C(H)(H) — Cl —[KOH (aq)]→ ....

d H — C(H)(H) — C(Cl)(H) — C(H)(H) — H —[KOH (aq)]→ ....

e H — C(H)(H) — C(Cl)(H) — C(H)(H) — H —[NH_3 (alc)]→ ....

EXAMINER'S TIP

You need to be able to link the type of reaction (i.e. addition, substitution ...) to the particular reaction occurring.

Example: Chloroethane + $NH_{3\,(Alc)}$ ⟶ Aminoethane

This reaction is a substitution reaction as the chloro is exchanged with an NH_2 group.

What do the following key words mean?

Elimination reaction	
Primary alcohol	
Secondary alcohol	
Tertiary alcohol	

ISBN: 9780170260107

Alcohols and haloalkanes

1 Complete the reaction scheme below of all the reactions ethanol can undergo.

Type of reaction: Oxidation

Name:

Structural formula:

Type of reaction:

Name: Chloroethane

Structural formula:

Type of reaction:

Name:

Structural formula:

```
  H       H
   \     /
H — C — C — H
   /     \
  H       Cl
```

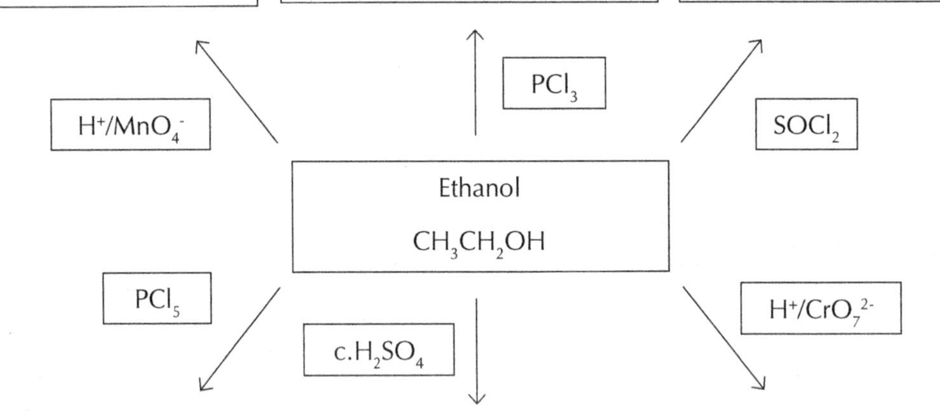

Type of reaction:

Name: Chloroethane

Structural formula:

Type of reaction: Elimination

Name:

Structural formula:

Type of reaction:

Name:

Structural formula:

```
    H   H   H
    |   |   |      O
H — C — C — C — C //
    |   |   |      \
    H   H   H       O — H
```


 ISBN: 9780170260107

2 Complete the reaction scheme, which shows some of the reactions bromoethane can undergo.

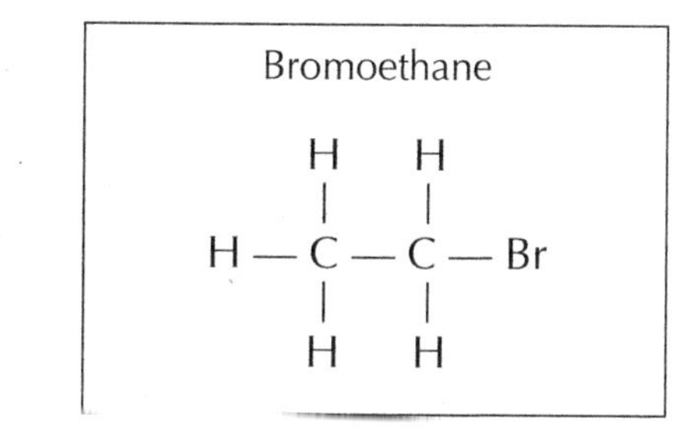

NH_3 (alc)

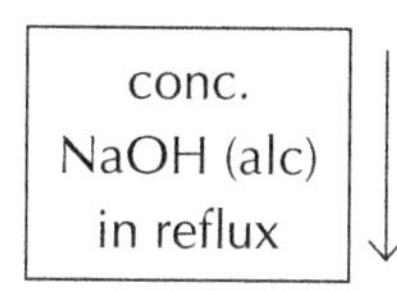

KOH (aq)

Type of reaction:	**Type of reaction:**	**Type of reaction:**
Name:	**Name:**	**Name:**
Structural formula:	**Structural formula:**	**Structural formula:**

3 Explain how you could tell the difference between two unlabelled bottles, one with 1-bromopropane the other with propan-1-ol using acidified dichromate.

In your answer include:

- Any reaction that would occur; with observations.
- The name of that reaction type.

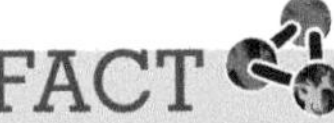

Carboxylic acids

Carboxylic acids are a functional group of *weak* acids, which have high melting points and boiling points due to the polar **COOH** end. They are soluble in water when they have a small carbon chain due to this polar end. However when the functional group is attached to a longer chain they lose their solubility in polar solvents.

FACT

Octanoic acid is found in the milk of many mammals. Goats' milk has three carboxylic acids in it making up 15% of its volume: octanoic acid, hexanoic acid and decanoic acid (that has 10 carbons in it).

Reactions of carboxylic acids

Dissociation in water

As carboxylic acids are weak acids they only partly dissociate in water to form their conjugate base and hydronium ion.

$CH_3COOH + H_2O \rightleftharpoons CH_3COO^- + H_3O^+$

Ethanoic acid + water $\rightleftharpoons$ ethanoate ion + hydronium ion

Neutralisation

Even though carboxylic acids are weak acids they will still react completely with a base or a carbonate to neutralise them. The reaction rate will be slower than a strong acid with a similar concentration.

$CH_3COOH + NaOH \longrightarrow NaCH_3COO + H_2O$

Ethanoic acid + sodium hydroxide $\longrightarrow$ Sodium ethanoate + water

$2CH_3COOH + Na_2CO_3 \longrightarrow 2NaCH_3COO + H_2O + CO_2$

Remember acid + carbonate reactions also produce carbon dioxide gas.

Metals

Carboxylic acids will also react with metals to produce hydrogen gas. Once again the reaction rate is slow due to their low dissociation rate.

$2CH_3COOH + Mg \longrightarrow Mg(CH_3COO)_2 + H_2$

2.13: Reactions of carboxylic acids

Complete and balance the following equations of reactions of carboxylic acids.

1 $CH_3CH_2COOH + H_2O \rightleftharpoons$ ____________________

2 $HCOOH + H_2O \rightleftharpoons$ ____________________

3 $CH_3CH_2COOH + Mg(OH)_2 \longrightarrow$ ____________________

4 $CH_3CH_2COOH + Li \longrightarrow$ ____________________

5 $HCOOH + Al_2(CO_3)_3 \longrightarrow$ ____________________

6 $HCOOH + Zn \longrightarrow$ ____________________

ISBN: 9780170260107

CHEMISTRY APPS

Aspirin is an example of a carboxylic acid that is used to treat pain, inflammation and fever. It was first isolated by Felix Hoffmann in 1897. Today aspirin is still one of the most widely used drugs with 40,000 tonnes of it being consumed each year.

Amines

Amines are characterised by their distinctive 'rotting fish' smell. In fact, amines are released when fish rot. They have a polar end, the NH_2 functional group, and so they have high melting points and boiling points. Just like carboxylic acids, the small-chain amines are soluble in water and other polar solvents, but are less soluble as the chain extends.

Reactions of amines

Amines are weak bases and so only partially dissociate in water and can be neutralised by acids.

Dissociation in water

When amines dissociate they produce hydroxide ions and their conjugate acid.

$CH_3NH_2 + H_2O \rightleftharpoons CH_3NH_3^+ + OH^-$

Methanamine + water ⇌ Methyl ammonium ion + hydroxide ion

Neutralisation

This reaction occurs slowly with an acid, whilst the equilibrium above is forced to go to completion.

$CH_3NH_2 + HCl \longrightarrow CH_3NH_3^+Cl^- + H_2O$

Methamine + hydrochloric acid ⟶ Methyl ammonium chloride + water

2.14: Reactions of amines

Complete the following equations showing reactions of amines.

1 $CH_3CH_2NH_2 + H_2O \rightleftharpoons$ ______________________________

2 $CH_3CH_2CH_2CH_2NH_2 + H_2O \rightleftharpoons$ ______________________________

3 $CH_3CH_2NH_2 + HCl \longrightarrow$ ______________________________

4 $CH_3CH_2NH_2 + HNO_3 \longrightarrow$ ______________________________

5 $CH_3CH_2NH_2 + H_2SO_4 \longrightarrow$ ______________________________

ISBN: 9780170260107

What do the following key words or phrases mean?

Neutralisation reaction	
Dissociation	

EXPERIMENT 3

Properties of carboxylic acids and amines

AIM: To investigate the properties of carboxylic acids and amines.

EQUIPMENT and CHEMICALS:

test tubes
magnesium ribbon pieces
ethanoic acid (2 mol L^{-1})
calcium carbonate chips
blue litmus paper
propylamine

SAFETY PRECAUTIONS:

Propylamine is a very hazardous skin, eye and lung irritant. Ethanoic acid is corrosive and is also a skin and eye irritant.

METHOD and RESULTS:

1 Properties of propylamine

	Observations		
	Blue litmus	Red litmus	Carefully smell by using the wafting technique
To 1 mL of propylamine			

 ISBN: 9780170260107

2 Properties and reactions of ethanoic acid

	Observations				
	Blue litmus	Red litmus	Carefully smell (by using the wafting technique)	Add a piece of Mg	Add a few calcium carbonate chips
To 1 mL of ethanoic acid					

QUESTIONS:

1 Write an equation for the reaction of ethanoic acid with magnesium.

2 Write an equation for the reaction of ethanoic acid with calcium carbonate.

3 Write an equation to show propylamine acting as a base.

4 Write an equation to show ethanoic acid acting as an acid.

ISBN: 9780170260107

CHECKPOINT 4

WHAT HAVE YOU LEARNED SO FAR?

Organic chemistry summary

Fill in the summary below by writing the type of reaction and the reagents required in the boxes below.

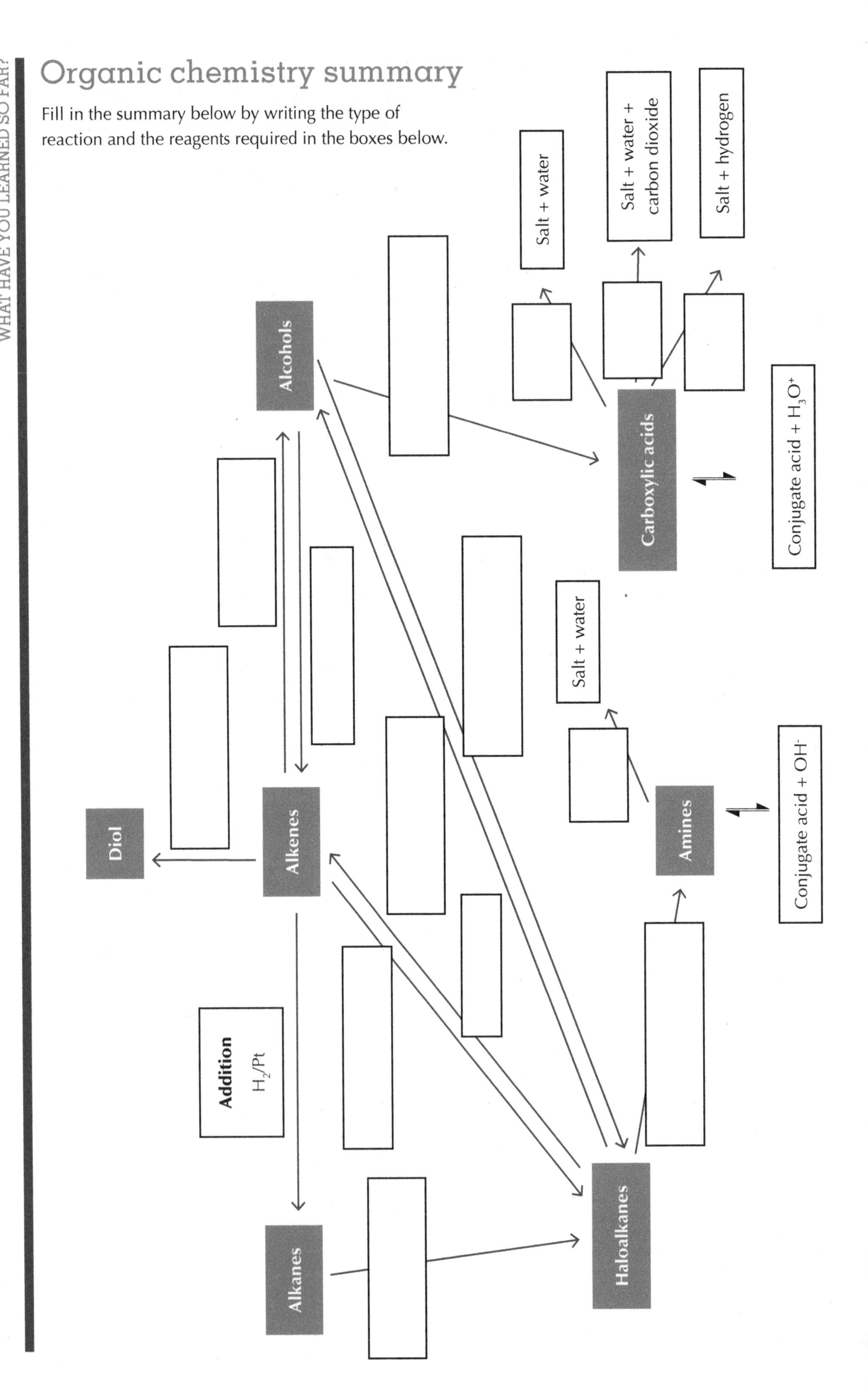

ISBN: 9780170260107

EXAM-TYPE QUESTIONS

Before you get started here are some points to remember from the Examiner's Report:
- Don't use generalisations, use terms carefully.
- Plan your answer before you start.

Question one

a Complete the table below, to demonstrate your understanding of organic structures and groups.

Substance	Structure	Name	Homologous series
A	$CH_3-CH(CH_3)-CH_2-OH$		
B	$CH_3-C(CH_3)(OH)-CH_3$		
C		1-aminobutane	
D		2-chloropropanoic acid	
E		methylpropene	

b Structural and geometric isomers are possible with a number of organic substances.

i Explain what structural isomers are, using C_4H_8 as an example.

ISBN: 9780170260107

ii Using C_4H_8 as an example, explain what conditions are necessary for geometric isomers to be possible.

Question two

a Complete the following table to show the structural formula and IUPAC (systematic) name for each compound.

Structural formula	Name
	2-methylbutanoic acid
$CH_3CH_2–NH_2$	
	2,3-dichlorobutane
$CH_2=C(CH_3)–CH_3$ (displayed formula: H₂C=C with CH₃ and CH₃ on second carbon)	

 ISBN: 9780170260107

Structural formula	Name
	2-methylbutan-2-ol

b State how you could tell the difference between aminopropane and propanoic acid using universal indicator.

c Propanoic acid neutralises sodium hydroxide but aminopropane does not. Explain why there is this difference between these two reactions. Describe how you could use universal indicator to find whether the sodium hydroxide has been neutralised.

Question three

a Complete the following table to show the IUPAC (systematic) name for each compound.

	Structural formula	Name
Molecule A	HC(Br)=CH(Br)	
Molecule B	$Br_2C=CH_2$	
Molecule C	$BrCH_2CH_2Br$	

b Explain why molecules A and B are structural isomers of each other, but molecule C is not.

ISBN: 9780170260107

c Explain why molecule A can form geometric isomers but molecule B cannot.

Question four

Two bottles containing propene ($CH_3CH=CH_2$), and propane ($CH_3CH_2CH_3$), require identification. Two reagents, bromine water (Br_2), and acidified permanganate (MnO_4^- / H^+), are available.

Evaluate the possible use of **BOTH** reagents to distinguish between the propene and propane. In your answer you should include:

- a description of the type of reactions that would occur
- any conditions that would be required
- any observations that would be made
- equations showing the structural formulas of the organic reactant(s) and product(s).

 ISBN: 9780170260107

Question five

a The molecule

$$\begin{array}{ccc} H & & H \\ & \diagdown \quad \diagup & \\ & C{=}C & \\ & \diagup \quad \diagdown & \\ H & & H \end{array}$$

undergoes an addition polymerisation reaction to form a polymer.

Draw the structure of **THREE** repeating units for the polymer made from this molecule.

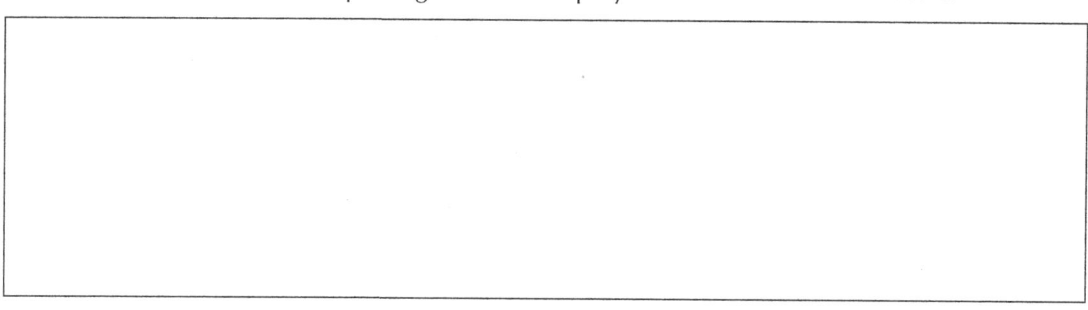

b For the following reaction:

$CH_2\,CH_2 + HBr \longrightarrow$

i Write in the box above the name and structural formula of the organic product formed.

ii Give reasons why this reaction can be classified as an addition reaction.

c Discuss why with the reaction above there is only one product, whereas with but-2-ene there would be two.

Include in your answer:

- A reaction of but-2-ene with hydrogen bromide, HBr.
- Which product would be the major product, and which would be the minor.
- Why ethene has only one product compared to the two formed with but-2-ene.

ISBN: 9780170260107

Question six

Pentan-1-ol can be oxidised to form a carboxylic acid.

i Write the name or formula of a suitable reagent that could be used to carry out the reaction. Include any specific conditions.

ii Describe the colour change that would be observed.

iii Write an equation to show this reaction.

Question seven

a State how you could distinguish between aminoethane ($CH_3CH_2NH_2$), and ethanoic acid (CH_3COOH), using damp litmus paper.

b When ethanoic acid reacts with magnesium carbonate ($MgCO_3$), fizzing can be seen during the reaction.

i What type of reaction is occurring?

ii Explain why fizzing occurs during the reaction.

iii Complete the equation below to show the structural formula of the organic product formed.

$$CH_3COOH + MgCO_3 \longrightarrow$$

iv Explain why this reaction has a slow rate of reaction.

ISBN: 9780170260107

EXAM-TYPE QUESTIONS

Question eight

Chloropropane, $CH_3CH_2CH_2Cl$, reacts with aqueous KOH, alcoholic NH_3 and concentrated NaOH (alc) under reflux.

Compare and contrast the reactions of chloropropane with these three reagents (KOH, NH_3, NaOH). In your answer you should include:

- the type of reaction occurring and the reason why it is classified as that type
- the type of functional group formed
- equations showing structural formulae for reactions occurring.

Question nine

Pent-1-ene is used in the reaction sequence shown below.

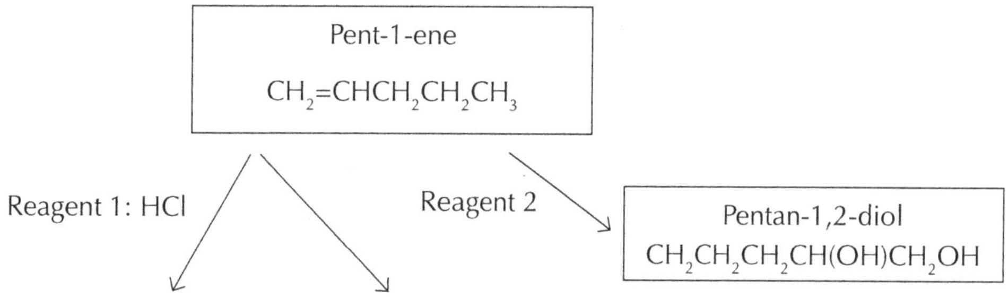

Pent-1-ene
$CH_2{=}CHCH_2CH_2CH_3$

Reagent 1: HCl

Reagent 2

Pentan-1,2-diol
$CH_3CH_2CH_2CH(OH)CH_2OH$

Name:	**Name:**
Structural formula:	**Structural formula:**
Major/Minor product	**Major/Minor product**

a **i** Name a suitable reagent for Reagent 2 shown at the bottom of page 103.

ii What type of reaction is occurring to make the pentan-1,2-diol?

b **i** Fill in the boxes on the previous page to show the organic products of the reaction between pent-1-ene and HCl.

ii What type of reaction is occurring to make the product from HCl reacting with pent-1-ene?

iii Circle the products you have drawn on page 103 as either major or minor products.

iv Explain why you made this choice using Markovnikov's rule.

Question ten

a Classify the haloalkanes below as primary, secondary or tertiary and give a reason why you made this choice for each.

```
         H
         |
      H— C —H
    H    |    H    H
    |    |    |    |
H — C —  C —  C —  C — H
    |    |    |    |
    H    Cl   H    H
```

Classification:

Reason:

```
         H
         |
      H— C —H
    H    |    H    H
    |    |    |    |
H — C —  C —  C —  C — H
    |    |    |    |
    H    H    Cl   H
```

Classification:

Reason:

 ISBN: 9780170260107

EXAM-TYPE QUESTIONS

b Explain why the molecules are structural isomers of each other.

c Draw as many other structural isomers as you can from this same molecular formula, $C_5H_{11}Cl$.

CHAPTER THREE

2.6 Demonstrate understanding of chemical reactivity (91166)

Learning outcomes

Tick off when you have studied this idea in class and when you have revised over that section prior to your assessment.

	Learning Outcomes	In class	Revision
1	Explain using collision theory how the rate of a reaction can be changed by changing the concentration.		
2	Explain using collision theory how the rate of a reaction can be changed by adding a catalyst.		
3	Explain using collision theory how the rate of a reaction can be changed by changing the temperature.		
4	Explain using collision theory how the rate of a reaction can be changed by the surface area.		
5	Explain how activation energy (E_A) may be altered.		
6	Explain what a dynamic equilibrium is.		
7	Write equilibrium constant (K_c) expressions for given equilibria.		
8	Explain how the position of an equilibrium can be affected and the affect on K_c with a change in concentration of either a reactant or product.		
9	Explain how the position of an equilibrium can be affected and the affect on K_c with a change in temperature.		
10	Explain how the position of an equilibrium can be affected and the affect on K_c with the addition of a catalyst.		
11	Explain how the position of an equilibrium can be affected and the affect on K_c with a change in pressure.		
12	Calculate K_c or changes to concentration of reactants or products.		
13	Define an acid and a base in terms of proton transfer.		
14	Define a strong, weak, concentrated and dilute acid.		
15	Explain how different strengths or concentrations of acids may vary conductivity, pH and reactivity.		
16	Calculate the pH of a strong acid.		
17	Use K_w to calculate the concentration or pH of a strong base.		

ISBN: 9780170260107

Pre-test – What do you know?

Section 1: Rates of reaction

1 In what four ways can you speed up the rate of a reaction?

2 What is collision theory? What does it have to do with rate of reaction?

Section 2: Reversible reactions

1 What does the word equilibrium mean to you?

2 Can you think of any reactions that are reversible? How do you know they are reversible?

Section 3: Acids and bases

1 What is an acid?

2 What is a base?

3 What happens when an acid reacts with a base?

4 What happens when an acid reacts with a carbonate?

5 What does pH mean?

6 What is an indicator? Name three or more.

Rates of reaction

The rate of a reaction depends upon the **collision theory of particles**. The collision theory of particles is that in order for reactants to react with each other they must *hit each other* with *enough force* and be in the *correct orientation* in order for the products to form.

We can speed up the rate of collisions in the following ways:

1 **Add a catalyst –** catalysts provide an alternative pathway for the reactants to react upon by lowering the activation energy (the energy required to make a reaction occur). They are not used up in the reaction, they just ensure that more reactants have enough energy to react to become the products.

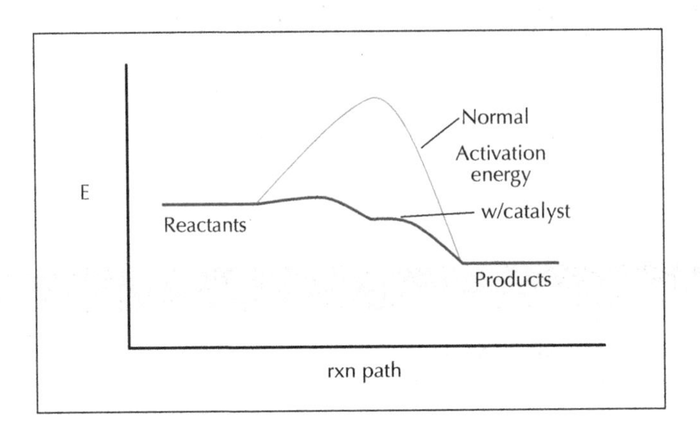

2 **Increasing the concentration –** by increasing the concentration of a solution of a reactant there will be more particles available in solution in order to react with, and therefore there will be more collisions per unit of time, therefore a faster rate of a reaction.

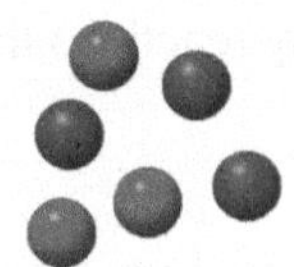

3 **Increasing the surface area –** by increasing the surface area there are more particles available to react with, and therefore there will be more collisions occurring per unit of time, and so a faster rate of a reaction.

There is a much larger surface area in the powder than the crystals, allowing more particles to be available to react.

4 **Increasing the temperature –** this increases the kinetic energy of the reactants, meaning they will collide more frequently and they will be more successful (collisions with enough force) per unit of time, and so a faster rate of a reaction.

 ISBN: 9780170260107

You will need to be able to explain the four mechanisms on page 108 as well as understand graphs like the one below.

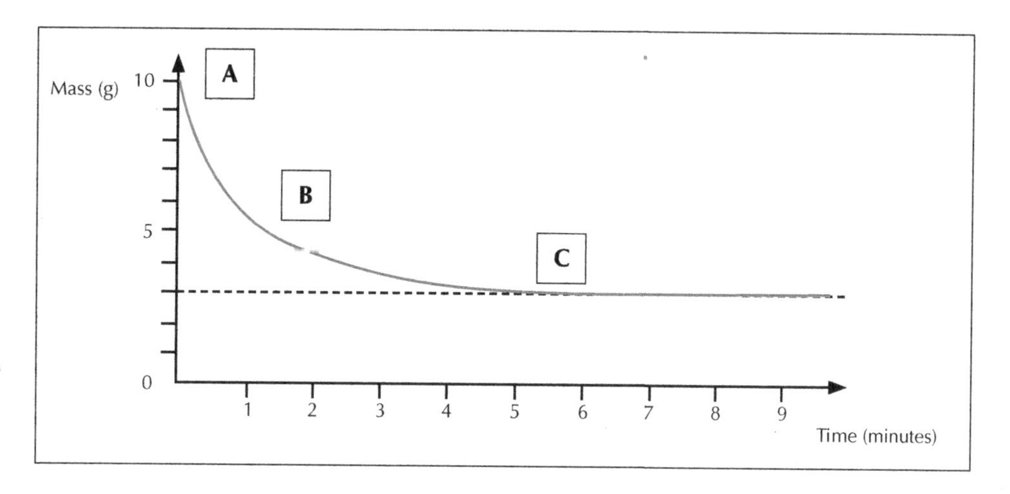

A In the beginning of the reaction there is a higher concentration of reactants hence a faster rate of reaction and a greater slope on the curve.

B In the second part of the curve the reaction rate slows down as there is a lower concentration of reactants to products because they have reacted together.

C In the final section the graph slopes to a horizontal line as the reaction is now complete, all the reactants have become the products or one of the reactants has been used up.

CHEMISTRY APPS

In 1971 a woman's body was discovered in an ancient coffin in Changsa, China. She was given the name Lady Dai, and was found to have been buried in 186 BC. She was overweight, had diabetes, high blood pressure, high cholesterol, and liver disease. We know this because the condition of the body was as if she had only been dead a few days. Inside the sealed coffin scientists found a liquid mixture consisting of mercury compounds, sulfur, and methane gas. How did her body remain so well preserved for so long? See if you can explain this puzzle in terms of chemistry and the rates of reaction.

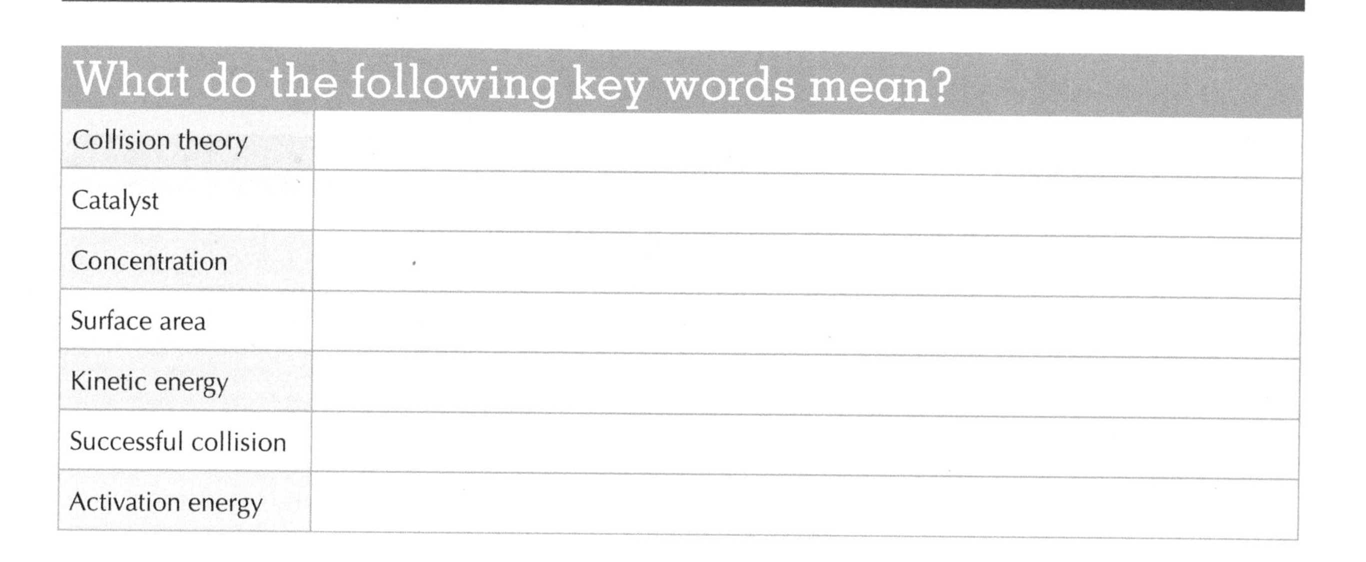

What do the following key words mean?

Collision theory	
Catalyst	
Concentration	
Surface area	
Kinetic energy	
Successful collision	
Activation energy	

ISBN: 9780170260107

3.1: Rates of reaction

1 Use collision theory to explain the following.

a Adding a little powdered manganese dioxide to hydrogen peroxide solution causes it to decompose and form oxygen very rapidly, but at the end of the reaction the manganese dioxide is unchanged.

b Marble chips (calcium carbonate) react much more slowly than powdered marble with hydrochloric acid.

c Calcium carbonate reacts much faster with 2 $molL^{-1}$ hydrochloric acid than with 0.2 $molL^{-1}$.

d Hydrogen peroxide is stored in the fridge to prevent it from decomposing.

EXAMINER'S TIP

1. When answering the questions above make sure you stated that the *number of collisions increased per second*, not just that the number of collisions increased.
2. Also when you talk about surface area make sure to say more particles are exposed for collisions to occur with.

ISBN: 9780170260107

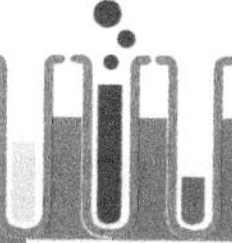

EXPERIMENT 1

How do catalysts alter the rate of reaction?

AIM: To demonstrate the effect of catalysts on the rate of reaction.

EQUIPMENT and CHEMICALS:

test tubes
sulfuric acid (2 mol L^{-1})
hydrogen peroxide (20%)
copper wire
manganese dioxide
stopwatch
granulated zinc

SAFETY PRECAUTIONS:

Hydrogen peroxide is an irritant to your eyes, skin and lungs. Manganese dioxide is an irritant to skin and eyes. Sulfuric acid is corrosive.

METHOD:

1 **The catalytic decomposition of hydrogen peroxide by manganese dioxide.**
 a Place 2 mL hydrogen peroxide in two separate test tubes.
 b To one test tube add a pinch of manganese dioxide.
 c Test the gas given off to identify it by placing a glowing splint into the test tube.

2 **The catalysis of the reaction between zinc and dilute sulfuric acid by copper.**
 a Place 3 mL sulfuric acid in two separate test tubes.
 b Place a piece of zinc in one test tube.
 c Place a piece of copper metal in contact with a piece of zinc in the second test tube.

RESULTS:

Reaction	Observations	Gas test
Hydrogen peroxide without manganese dioxide		
Hydrogen peroxide with manganese dioxide		

Reaction	Observations
Zinc and sulfuric acid without copper	
Zinc and sulfuric acid with copper	

QUESTIONS:

1 The balanced equation for hydrogen peroxide decomposing is $2H_2O_2 \xrightarrow{MnO_2} O_2 + 2H_2O$. How could you prove that the manganese dioxide was not used up during the reaction?

2 How does copper affect the rate of reaction between zinc and sulfuric acid?

3 Write a balanced equation for the reaction occurring between Zn and H_2SO_4.

ISBN: 9780170260107

CHEMISTRY APPS

Enzymes are biological catalysts. They work by having a specific place in their structure to bring together two chemicals. These chemicals then react. Most of the reactions in your body would not occur without enzymes as these reactions often have such a large activation energy.

For example, this large molecule is the enzyme used in PCR (polymerase chain reactions) machines in order to copy DNA from a sample so that the DNA sample can be studied.

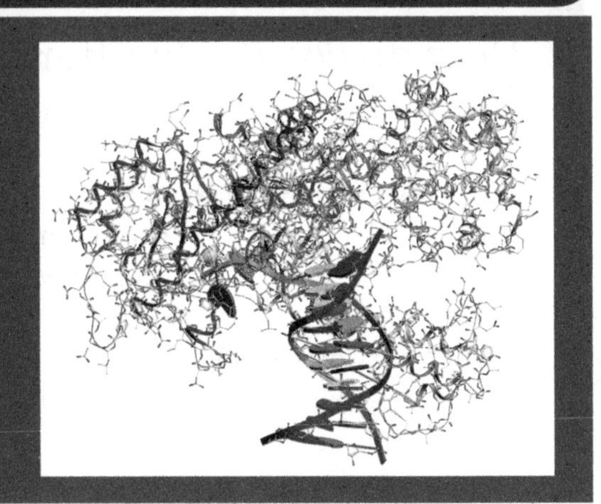

INVESTIGATION 1

How does concentration affect the rate of a reaction?

AIM: To investigate how a change in concentration affects the rate of a reaction.

EQUIPMENT and CHEMICALS:

test tubes
250 mL beaker
50 mL beaker
10 mL measuring cylinder
hydrochloric acid (2 mol L^{-1})
magnesium ribbon
stopwatch

METHOD:

Describe your own method to find how the concentration of acid used affects the rate of how fast the magnesium metal disappears. You need to choose at least three different concentrations of the acid and ensure you keep all other variables the same.

 ISBN: 9780170260107

RESULTS:

Draw a table of your results here.

QUESTIONS:

1 As the concentration of the acid is decreased, what happens to the time taken for the magnesium metal to disappear?

2 Write a balanced equation for the reaction:

3 What did you notice about the rate of the reaction as the reaction proceeded? Was there a difference between the start and the end?

ISBN: 9780170260107

INVESTIGATION 2

How does the surface area affect the rate of a reaction?

AIM: To investigate the effect of surface area on the rate of a reaction.

EQUIPMENT AND CHEMICALS:

test tubes
hydrochloric acid (2 mol L^{-1})
calcium carbonate chips
calcium carbonate powder
stopwatch

METHOD:

Describe your own method to find how the surface area of calcium carbonate used affects the rate of reaction. Ensure you keep all other variables the same.

RESULTS:

Draw a table of your results here.

ISBN: 9780170260107

QUESTIONS:

1 Which form of calcium carbonate had a faster rate of reaction? Why?

2 How could you have improved your method?

3 Give an example from everyday life where increasing the surface area affects the rate of a reaction.

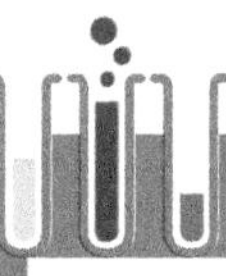

EXPERIMENT 2

Temperature and rate of reaction

AIM: To find out if temperature affects the rate of a reaction.

EQUIPMENT:

1 $molL^{-1}$ potassium thiosulfate solution
1 $molL^{-1}$ hydrochloric acid
100 mL beaker
Bunsen burner
thermometer
100 mL measuring cylinder
stopwatch

METHOD:

1. Measure 20 mL of potassium thiosulfate and pour into a 100 mL beaker.
2. Measure 20 mL of hydrochloric acid and add to the same beaker.
3. Place the beaker over a cross drawn on a piece of paper, time until you can't see the cross any more.
4. Heat up 20 mL of potassium thiosulfate over a Bunsen till it heats up to 30 °C.
5. Remove from the heat and add 20 mL of hydrochloric acid and place over the cross again, time until you can't see it any more.
6. Now heat up 20 mL of potassium thiosulfate over a Bunsen burner until it heats up to 40 °C.
7. Remove from the heat and add 20 mL of hydrochloric acid and place over the cross again, time until you can't see it any more.
8. Now heat up 20 mL of potassium thiosulfate over a Bunsen burner until it heaps up to 50 °C. Remove from the heat and add 20 mL of hydrochloric acid and place over the cross again, time until you can't see it any more.

ISBN: 9780170260107

RESULTS:

Temperature (°C)	Time taken (seconds)
Room temperature	
30	
40	
50	

CONCLUSION:

As the temperature is increased the rate of reaction ______. This is because the particles are moving ______ in solution, therefore there are more ______ occurring. This ______ the rate of reaction. This means more particles have enough activation energy to react.

QUESTIONS:

1 As the temperature increases, what happens to the time taken for the reaction to occur?

2 Describe how you could control the rate of a reaction which was exothermic.

 ISBN: 9780170260107

CHECKPOINT 1
WHAT HAVE YOU LEARNED SO FAR?

Rate of reaction

1 Describe how you would go about measuring the rate of the reaction between magnesium metal and hydrochloric acid. (What measurements would you make?)

2 In a reaction between hydrochloric acid and zinc metal, a lump of zinc weighed 2 g before it was placed in the acid. It was carefully removed at regular time intervals and reweighed. The mass versus time graph of the zinc is shown below.

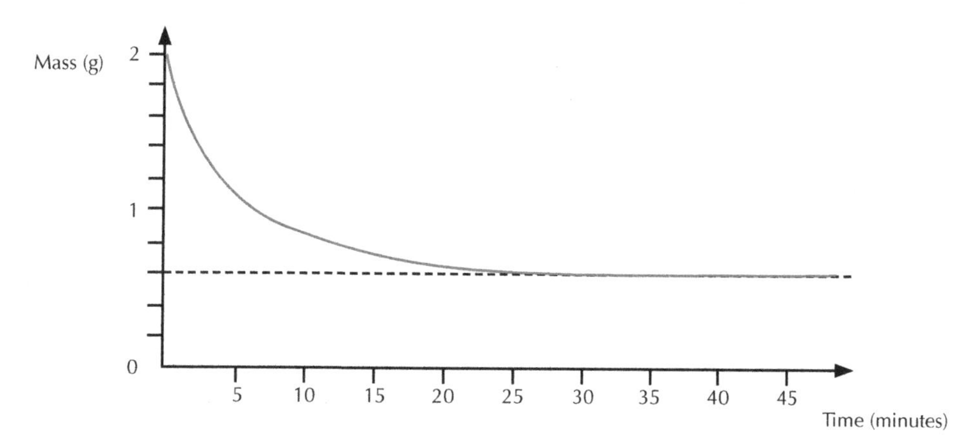

a How much zinc had reacted when the reaction had finished?

b After how many minutes did the reaction appear to have finished?

c State when the reaction is fastest and explain how you determined this.

3 What are the two requirements for a collision between two reacting molecules to be effective and lead to a reaction?

ISBN: 9780170260107

Reversible reactions

Some chemical reactions can be reversed because not all the reactants are used up as the reaction proceeds. For example, we can reverse the reaction between hydrogen and nitrogen to make ammonia. Instead of drawing a complete arrow ($\longrightarrow$) we would draw a reversible arrow like this $\rightleftharpoons$ to show the reaction is occurring in both directions.

$3H_{2\,(g)} + N_{2\,(g)} \rightleftharpoons 2NH_{3\,(g)}$

Note the (g) next to the reactants and products tells us what state they are in, which is important as a reaction will be different if the reactants and products are in different states.

> (**g**) stands for gas, but we can also have (**s**) for solid or (**l**) for liquid and finally (**aq**) for dissolved in water or aqueous.

Equilibrium

How do we know how much product there will be? The answer: reversible reactions have an equilibrium point where the rate at which reactants change into products is equal to the rate at which products change into reactants. This is not to say there will be equal amounts of both reactants and products, as equilibria usually tend either towards the reactants or products. If you think of walking up a 'down' escalator, the point where neither you nor the escalator is gaining on the other, because you are both going at the same speed, is like a chemical equilibrium. As equilibria like this are constantly moving they don't have a fixed end point so we call them **dynamic**.

What do the following key words or phrases mean?

Reversible reaction	
Aqueous	
Chemical equilibrium	
Dynamic equilibrium	

3.2: Reversible reactions

1 How would you know if a reaction was reversible?

2 Is iron rusting a reversible reaction? How do you know?

 ISBN: 9780170260107

3 Is burning toast a reversible reaction? How do you know?

4 Is melting ice a reversible reaction? How do you know?

5 Name one reversible reaction.

CHEMISTRY APPS

One important reversible reaction occurring in your body is between the oxygen you breathe in for respiration and the haemoglobin in your blood. Haemoglobin is a large molecule which binds to the oxygen to transport it all around your body to your cells where the oxygen unbinds itself in order for your cells to break down glucose into a useable energy source.

The equilibrium constant, K_c

Dynamic equilibria have a specific value (K_c, the equilibrium constant of concentration) and are calculated from the concentration of the products divided by the concentration of the reactants.

$$K_c = \frac{[\text{products}]}{[\text{reactants}]}$$

Note [ ] means the concentration of in $molL^{-1}$

For any equilibrium at any temperature and pressure there is a K_c value or number that tells you where an equilibrium is placed. In the example below you will notice that the concentration of ammonia $[NH_3]$ is squared and the concentration of hydrogen $[H_2]$ is cubed. When writing the equilibrium constant expression for any given reaction you should always place the number of moles of that reactant or product to the power of that concentration. This reaction shows the Haber process, used for making NH_3.

$$3H_{2\,(g)} + N_{2\,(g)} \rightleftharpoons 2NH_{3\,(g)} \text{ at } 300\ °C \qquad K_c = 4.34 \times 10^{-3} = \frac{[NH_3]^2}{[N_2][H_2]^3}$$

Finally you should not place in reactants or products that are in solid or liquid states as their concentrations generally do not alter much in an equilibrium.

$$CuO_{(s)} + CO_{(g)} \rightleftharpoons Cu_{(s)} + CO_{2(g)} \qquad K_c = \frac{[CO_2]}{[CO]}$$

If K_c is a large value (over 1) then we know the reaction is strongly product-favoured (because the products are on the top of the fraction). If K_c is small (less than 1) then we know the reaction is reactant-favoured. So at 300 °C the above reaction (the Haber process) is reactant-favoured and at the equilibrium the concentration of the reactants nitrogen and hydrogen is larger than the concentration of the product ammonia.

You will need to write equilibrium constant expressions for any given equilibrium.

ISBN: 9780170260107

What do the following key words or phrases mean?

Equilibrium constant	
Concentration	

3.3: Equilibrium constants, K_c

Write the equilibrium constant expressions using the following information, note the first two have been done for you to help you.

1	$PCl_{3(g)} + Cl_{2(g)} \rightleftharpoons PCl_{5(g)}$	$K_c = \frac{[PCl_5]}{[PCl_3][Cl_2]}$
2	$2N_2O_{5(g)} \rightleftharpoons 4NO_{2(g)} + O_{2(g)}$	$K_c = \frac{[NO_2]^4[O_2]}{[N_2O_5]^2}$
3	$SO_3^{2-}{}_{(aq)} + H_2O_{(l)} \rightleftharpoons HSO_3^-{}_{(aq)} + OH^-{}_{(aq)}$	
4	$CO_{(g)} + H_2O_{(g)} \rightleftharpoons CO_{2(g)} + H_{2(g)}$	
5	$Fe^{2+}{}_{(aq)} + Ag^+{}_{(aq)} \rightleftharpoons Ag_{(s)} + Fe^{3+}{}_{(aq)}$	
6	$2CO_{(g)} + O_{2(g)} \rightleftharpoons 2CO_{2(g)}$	
7	$3O_{2(g)} \rightleftharpoons 2O_{3(g)}$	
8	$Cu(H_2O)_6^{2+}{}_{(aq)} + 4Cl^-{}_{(aq)} \rightleftharpoons CuCl_4^{2-}{}_{(aq)} + 4H_2O_{(l)}$	
9	$2C_{(s)} + O_{2(g)} \rightleftharpoons 2CO_{(g)}$	
10	$4NH_{3(g)} + 5O_{2(g)} \rightleftharpoons 4NO_{(g)} + 6H_2O_{(g)}$	

 ISBN: 9780170260107

Changes to equilibria

We can alter the position of an equilibrium (K) by doing one of the following things:

1 **Concentration** - adding more of a reactant or product
2 **Pressure** - increasing or decreasing the pressure
3 **Temperature** – increasing or decreasing the temperature
4 **Catalyst** – this speeds up the rate of reaction in both the forward and reverse directions equally making no change to an equilibrium

Le Chatelier's principle

In order to explain how the above four things alter an equilibrium we have to understand Le Chatelier's principle:

Any change made to an equilibrium system will make an equilibrium shift in order to correct that change.

What the principle means is if we change the concentration, pressure or temperature of an equilibrium system the equilibrium position must move in order to correct that change. It will either move in the forward direction of a reaction (towards the products) or in the backward direction (towards the reactants).

Note that all the equilibria we talk about in this topic are assumed to be closed systems, meaning that none of the reactants or products can escape as this would force an equilibrium towards completion if that were the case.

For example, fizzy drinks are in equilibrium when properly sealed in their bottles with the carbon dioxide both dissolving in and out of the solution. Once opened the carbon dioxide leaks out leaving a flat solution with no carbon dioxide left.

FACT

Henry Louis Le Chatelier first presented his principle in 1885 and wrote many papers on this topic. He originally trained as an engineer, however he decided he would prefer to teach Chemistry which he did at École des Mines in Paris.

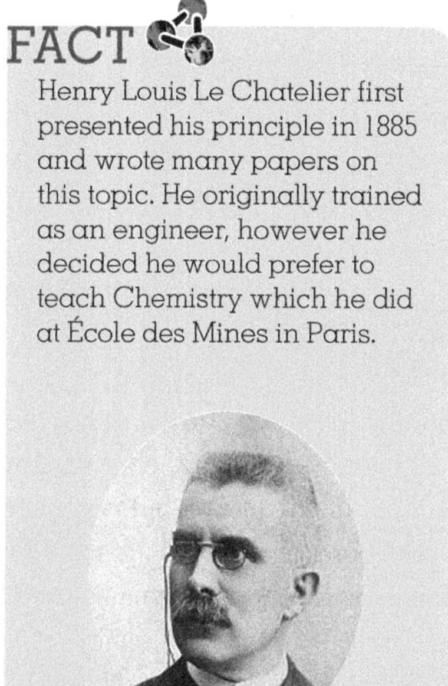

Changing concentration

So, for example, if we add more of a reactant, either nitrogen or hydrogen in this case, to the system below, the equilibrium will shift towards the products in order to correct that change. This will mean we have more ammonia (NH_3) in our equilibrium than before.

$$N_{2\,(g)} + 3H_{2\,(g)} \rightleftharpoons 2NH_{3\,(g)}$$

However if we add more ammonia then the equilibrium will shift towards the reactants. This will mean we have more of both nitrogen and hydrogen in our equilibrium.

Neither of these changes will affect the value of K_c as all the values will change by equal amounts.

Changing pressure

If we increase the pressure or decrease the volume of the space where the above reaction is taking place the equilibrium will shift towards the side with the fewest moles of gas.

$$N_{2\,(g)} + 3H_{2\,(g)} \rightleftharpoons 2NH_{3\,(g)}$$

ISBN: 9780170260107

In the Haber process the equilibrium will shift towards the product ammonia (NH_3) as there are only two moles of gas compared to the four total on the reactant side.

However, if we increase the volume or decrease the pressure the equilibrium will shift to the side with the most moles of gas, in this case towards the reactants nitrogen and hydrogen.

Note: pressure will only affect equilibria with gases in them as gases can be compressed. Once again the K_c value will not be altered as there will be no overall change to the ratio of the products to the reactants.

Changing temperature

In reversible reactions there is always an endothermic direction and an exothermic direction. If you increase the temperature, then the equilibrium will always shift in the endothermic direction in order to minimise the change made by the extra heat.

On the other hand, if you decrease the temperature, then in order to minimise the change made to the system, the equilibrium will shift in the exothermic direction.

With the Haber process the forward direction (towards the product ammonia, NH_3) is exothermic and towards the reactants (nitrogen and hydrogen) is endothermic. So if we increase the temperature the equilibrium will shift towards the reactants.

K_c is affected by temperature and you will get specific K_c values for specific temperatures. So, for the Haber process,when we increase the temperature we will get a smaller K_c value since we will get more reactants. Whereas if we decrease the temperature we will get a larger K_c since there will be more products.

Catalysts

Finally, when we add a catalyst there is no change to an equilibrium position or value as the rate of reaction will increase in both the forward and reverse reactions equally. This means equilibrium will be reached faster but not altered.

All sorts of catalysts are used to increase the rate of reaction in the Haber process. Some examples of chemicals used are K_2O, CaO, SiO_2, and Al_2O_3.

FACT

In 1909 Fritz Haber of Germany discovered how to make ammonia (NH_3) from hydrogen gas and nitrogen gas, a process that was named after him. The reaction may sound simple, but Haber needed a good understanding of the 'new' Le Chatelier principle before he realised that the reaction could only work under enormous pressure. The discovery helped start the production of cheap nitrogen-rich fertilisers, which later became of great benefit to the whole world. Haber then used other ammonium compounds to develop explosives used in World War I. Clara Haber, his wife and also a talented chemist, was deeply distressed by his involvement in weapons production.

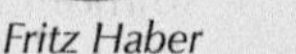
Fritz Haber

Clara Haber

What do the following key words or phrases mean?

Le Chatelier's principle	
Pressure	
Temperature	
Catalyst	

ISBN: 9780170260107

3.4: Le Chatelier's principle

Explain what effect (if any) the following changes would have on equilibria 1 to 5 below. Use Le Chatelier's principle and any effect to the K_c value to explain your answer.

1 $4NH_{3(g)} + 5O_{2(g)} \rightleftharpoons 4NO_{(g)} + 6H_2O_{(g)}$ $\Delta_rH° = -46.2 \text{ kJ mol}^{-1}$

a Increasing the temperature.

b Decreasing the total pressure.

c Adding a catalyst.

d Decreasing the ammonia, NH_3 concentration.

2 $2SO_{2(g)} + O_{2(g)} \rightleftharpoons 2SO_{3(g)}$ $\Delta_rH° = -94.8 \text{ kJ mol}^{-1}$

a Increasing pressure.

b Adding more sulfur dioxide, SO_2.

c Adding a catalyst.

d Decreasing the temperature.

3 $Cu(H_2O)_6{}^{2+}{}_{(aq)} + 4Cl^-{}_{(aq)} \rightleftharpoons CuCl_4{}^{2-}{}_{(aq)} + 4H_2O_{(e)}$

$Cu(H_2O)_6{}^{2+}{}_{(aq)}$ is blue Cl^- is colourless $CuCl_4{}^{2-}$ is yellow

a What will happen to the colour of the system when more $Cu(H_2O)_6{}^{2+}$ is added?

b What will happen to the colour of the system when more Cl^- is added?

ISBN: 9780170260107

4 $CO_{(g)} + 2H_{2(g)} \rightleftharpoons CH_3OH_{(g)}$

What will happen to the amount of methanol, CH_3OH formed when the following changes are made to the equilibrium system?

a Increasing the pressure.

b Increasing the concentration of carbon monoxide, CO.

c Adding a catalyst.

5 $H_{2(g)} + I_{2(g)} \rightleftharpoons 2HI_{(g)}$ $\Delta_r H° = + 26.5\ kJmol^{-1}$

H_2 is colourless I_2 is purple HI is colourless

a What will happen to the amount of purple colour in the system when the temperature is increased?

b What will happen to the amount of hydrogen Iodide, HI produced when the temperature is decreased?

ISBN: 9780170260107

EXPERIMENT 3

Observing changes to a chemical equilibrium

AIM: To observe a chemical equilibrium using the Fe^{3+} (aq) / SCN^{-} (aq) system and see how it can be altered.

EQUIPMENT and CHEMICALS:

test tubes
potassium thiocyanate (0.1 mol L^{-1})
test tube rack
iron (III) nitrate (0.1 mol L^{-1})
sodium fluoride solid
potassium thiocyanate solid

SAFETY PRECAUTIONS:

Sodium fluoride is an irritant to skin and eyes and is also corrosive to these body parts. Potassium thiocyanate is also an irritant to eyes and skin as well as lungs. Iron (III) nitrate is a strong irritant to skin and eyes. ***Be sure to wash your hands after this practical!***

METHOD:

1 To 4 test tubes add 1 mL of potassium thiocyanate and then do the following:
  - **a** Leave one untouched. This is the control, so you can compare its colour with the others.
  - **b** To one add a 1 mL of iron (III) nitrate solution.
  - **c** To one add a 1 mL of iron (III) nitrate solution and then add a few crystals of potassium thiocyanate.
  - **d** To one add a 1 mL of iron (III) nitrate solution and then add a little sodium fluoride. (Fluoride ions react with the iron (III) ions.)

RESULTS and QUESTIONS:

1 What colour was the potassium thiocyanate solution?

____

2 What colour was the iron (III) nitrate solution?

____

3 What colour was the iron thiocyanate solution (the solution when you added the above two chemicals together)?

____

4 Complete the equilibrium equation for the reaction that occured:

__________ + __________ $\rightleftharpoons$ $FeSCN^{2+}_{(aq)}$

5 Explain what happened to the equilibrium system when more potassium thiocyanate was added.

____

____

____

____

____

____

____

____

____

ISBN: 9780170260107

6 Explain what happened to the equilibrium system when you removed some of the iron (III) ions by adding the sodium fluoride.

INVESTIGATION 3

Explaining the changes in concentration made to the chromate/dichromate equilibrium

AIM: Making observations and writing an equilibrium investigation on concentration.

EQUIPMENT and CHEMICALS:

test tubes
hydrochloric acid (2 mol L^{-1})
test tube rack
hydrochloric acid (1 mol L^{-1})
potassium chromate (0.1 mol L^{-1})
potassium dichromate (0.1 mol L^{-1})

SAFETY PRECAUTIONS:

Potassium chromate and potassium dichromate are both corrosive and skin, eye and lung irritants. Hydrochloric acid is corrosive.

METHOD:

You have been given all the chemicals involved in the equilibrium shown below: potassium chromate K_2CrO_4, two concentrations of hydrochloric acid, HCl which will give you the H^+ and potassium dichromate $K_2Cr_2O_7$ (note that the potassium ions and chloride ions are both spectators in this equilibrium which means we can ignore them for the moment). Write a method to show how you could test for changes made to this equilibrium by varying the concentration of hydrochloric acid added.

$$2\,CrO_4^{2-}{}_{(aq)} + 2H^+{}_{(aq)} \rightleftharpoons Cr_2O_7^{2-}{}_{(aq)} + H_2O_{(l)}$$

ISBN: 9780170260107

RESULTS:

Draw up a table to show what you found out.

CONCLUSION:

1 What happened to the colour of the chromate/dichromate equilibrium system when you increased the concentration of one of the reactants?

2 Explain this change in colour using Le Chatelier's principle.

3 Would this change affect the value of K_c? Why? Why not?

INVESTIGATION 4

Explaining the changes in temperature made to the cobalt ion equilibrium

AIM: Making observations and writing an equilibrium investigation on temperature.

EQUIPMENT and CHEMICALS:

cobalt chloride (0.1 mol L^{-1})
test tube rack
100 mL beaker
sodium chloride solid
Bunsen burner
thermometer
test tubes
tripod and gauze

SAFETY PRECAUTIONS:

Cobalt chloride is a strong irritant to skin, eyes and intestinal tract.

METHOD:

In order to make up the equilibrium below you will need to add 3 mL of cobalt chloride solution to a test tube then add in sodium chloride solid until no more will dissolve. Decant (pour off) the solution into a new test tube, this is the equilibrium below.

$$[Co(H_2O)_6]^{2+}{}_{(aq)} + 4Cl^-{}_{(aq)} \rightleftharpoons [CoCl_4]^{2-}{}_{(aq)} + 6H_2O_{(l)}$$

ISBN: 9780170260107

Write a method for how you could alter the temperature that the cobalt ion equilibrium is in. Now do the investigation. Use at least four different temperatures.

RESULTS:

Draw up a results table to show what you have found out.

CONCLUSION:

1 What happens to the cobalt ion equilibrium when you increase the temperature of the mixture?

2 Explain using Le Chatelier's principle how the increase in temperature affects the equilibrium.

3 Does this change alter the value of K_c? Why or why not?

ISBN: 9780170260107

4 What does this tell you about the enthalpy value of the forward reaction: is it positive or negative? Is the forward reaction endothermic or exothermic? Why?

3.5: NCEA-style Le Chatelier's principle questions

There are three key elements to answer these questions at Excellence level:

1 State the direction in which the equilibrium will shift.
2 Explain the shift relating back to Le Chatelier's principle.
3 Explain whether the value of K_c will change and how it will be changed, if it is altered.

For example:
Discuss the changes made to the Haber process equilibrium when the temperature of the system is increased.

$$N_{2\,(g)} + 3H_{2\,(g)} \rightleftharpoons 2NH_{3\,(g)} \qquad \Delta_r H° = \text{-ve kJ mol}^{-1} \qquad K_c = \frac{[NH_3]^2}{[N_2][H_2]^3}$$

The equilibrium will shift towards the reactants in order to minimise the change made to the system by the increase in temperature. This is so the equilibrium shifts in the endothermic direction in order to remove the extra heat added to the system. K_c will be altered and become smaller as there will be a greater concentration of reactants to products due to this change.

EXAMINER'S TIP

1 Remember K_c is only affected by a change in temperature.
2 When stating the shift, always say 'towards the reactants' or 'towards the products' rather than the less specific 'to the left or right' so the examiner is sure you know what you are talking about.

1 a Discuss the changes in the sulfur trioxide equilibrium when the temperature is decreased.

$$2\,SO_{2(g)} + O_{2(g)} \rightleftharpoons 2\,SO_{3(g)} \qquad \Delta_r H° = -197\text{ kJ mol}^{-1}$$

ISBN: 9780170260107

b Discuss the changes in the sulfur trioxide equilibrium above when the concentration of sulfur dioxide is halved.

2 a Discuss the changes in the carbon monoxide equilibrium when the pressure is increased.

$$H_2O_{(g)} + C_{(s)} \rightleftharpoons H_{2\,(g)} + CO_{(g)}$$

b Discuss the changes in the carbon monoxide equilibrium when a catalyst is added.

3 a Discuss the changes in the silver equilibrium when the concentration of copper ions is doubled.

$$Cu_{(s)} + Ag^{+}_{(aq)} \rightleftharpoons Cu^{2+}_{(aq)} + Ag_{(s)} \qquad \Delta_rH° = -147 \text{ kJ mol}^{-1}$$

b Discuss the changes in the silver equilibrium above when the temperature is increased.

Calculations involving K_c

If we know the concentration of the products and the reactants we can calculate the K_c value or any of the values in a K_c expression if required.

For example:

Calculate the value of K_c for the nitrogen dioxide equilibrium below:

$$2NO_{2\,(g)} \rightleftharpoons N_2O_{4\,(g)}$$

given that the concentration of NO_2 is 0.0251 $molL^{-1}$ and the concentration of N_2O_4 is 0.0869 $molL^{-1}$.

Step 1: Write the equilibrium constant, K_c expression for this equilibrium.

$$K_c = \frac{[N_2O_4]}{[NO_2]^2}$$

(Remember it is always [products] divided by [reactants])

Step 2: Put your values in and calculate.

$$K_c = \frac{[0.0869]}{[0.0251]^2} = 1.38 \text{ (3 s.f.)}$$

(Remember you don't have to write units for K_c)

3.6: K_c calculations

1 The equilibrium $2NO_{2(g)} \rightleftharpoons N_2O_{4(g)}$ was established at 25 °C.

The equilibrium mixture gave the concentration of N_2O_4 as 0.0201 $molL^{-1}$ and of NO_2 as 0.102 $molL^{-1}$. Determine the value of K_c under these conditions.

2 Analysis of the equilibrium system $2SO_{2(g)} + O_{2(g)} \rightleftharpoons 2SO_{3(g)}$ showed that: $[SO_2] = 0.240$ $molL^{-1}$, $[O_2] = 1.28$ $molL^{-1}$ and $[SO_3] = 0.931$ $molL^{-1}$.

Calculate the K_c value.

 ISBN: 9780170260107

3 Hydrogen and iodine are pumped into a 1L sealed container. The equilibrium amounts are: $n(H_2) = 0.0275$ mol, $n(I_2) = 0.0290$ mol and $n(HI) = 0.175$ mol.

The equation for the reaction is $H_{2(g)} + I_{2(g)} \rightleftharpoons 2HI_{(g)}$

Calculate the K_c value.

4 At 1200 °C, $K_c = 2.25 \times 10^{-4}$ for the equilibrium

$$2H_2S_{(g)} \rightleftharpoons 2H_{2(g)} + S_{2(g)}$$

In an equilibrium mixture $[H_2S] = 5.01 \times 10^{-3}$ $molL^{-1}$ and $[S_2] = 2.20 \times 10^{-3}$ $molL^{-1}$. Calculate the equilibrium concentration of $H_{2(g)}$.

CHECKPOINT 2

WHAT HAVE YOU LEARNED SO FAR?

Equilibrium systems

1 Colourless hydrogen and purple iodine gas react together to form colourless hydrogen iodide gas. The equilibrium is shown below:

$$H_{2\,(g)} + I_{2\,(g)} \rightleftharpoons 2HI_{\,(g)}$$

a Write the equilibrium expression, K_c for this reaction.

$K_c =$

b If K_c is 64 and the concentrations of hydrogen and iodine gases are both 0.100 $molL^{-1}$, what is the concentration of hydrogen iodide gas formed?

ISBN: 9780170260107

c With a K_c value of 64 which species is going to be present at a higher concentration: H_2 or HI. Explain why?

d If the $\Delta_r H°$ for this reaction is + 26.5 $kJmol^{-1}$, what would happen to the value of K_c and the position of the equilibrium if the temperature of the system was doubled?

e Discuss the effect of doubling the pressure of the system.

f Discuss the effect of halving the concentration of hydrogen gas present in the system.

Acids and bases

Acids and bases were originally defined by their properties such as *acids taste sour* and *bases taste bitter*, however like most things in chemistry the important thing is how they react.

Brønsted-Lowry definition

The Brønsted-Lowry definition for an **acid** is a *substance that donates a proton or an H^+;* and a **base** is *a substance that accepts a proton.* These are the definitions we will be using to help write reactions of acids and bases and explain how they react.

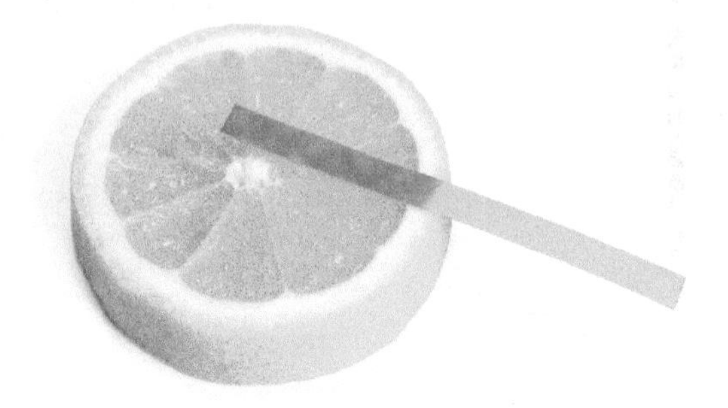

  ISBN: 9780170260107

First though, we need to know the difference between the words *strong/weak* and *dilute/concentrated.*

A strong acid is one which completely dissociates or ionises in water to form its conjugate base (what is left over from the acid when the H^+ is removed).

$$HCl + H_2O \longrightarrow Cl^- + H_3O^+$$

Note the reaction goes to completion. It is not reversible. The H^+ bonds to water to form the acid particle, also known as the *hydronium ion,* H_3O^+.

A strong base also completely dissociates in water to form its conjugate acid and hydroxide ions, OH^-.

$$NaOH \longrightarrow Na^+ + OH^-$$

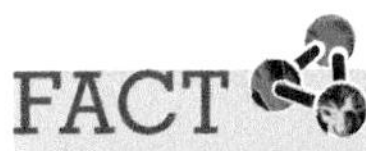

FACT

The strongest Brønsted-Lowry acid is fluorosulfonic acid, HSO_3F, which is more than 1000 times stronger than sulfuric acid, H_2SO_4.

Note that an acid will only show acidic properties like sourness or corrosiveness when in the presence of water because what causes these properties is the hydronium ion, H_3O^+.

A weak acid is one that only partially dissociates in water, forming its conjugate base and water. It is an equilibrium, like ethanoic acid (vinegar) shown below.

$$CH_3COOH + H_2O \rightleftharpoons CH_3COO^- + H_3O^+$$

A weak base also only partially dissociates in water to form its conjugate acid and hydroxide ions (from removing a hydrogen ion from water).

$$NH_3 + H_2O \rightleftharpoons NH_4^+ + OH^-$$

A concentrated acid is one which has a large number of acid particles in it (it could be a strong or weak acid). A dilute acid is one with few acid particles in it and it could be made from a concentrated or a dilute acid.

Other reactions

There are three other acid and base reactions you should be familiar with:

1 acid + base $\longrightarrow$ salt + water
   e.g. $HCl + NaOH \longrightarrow NaCl + H_2O$

2 acid + metal $\longrightarrow$ hydrogen + salt
   e.g. $2HCl + Mg \longrightarrow H_2 + MgCl_2$

3 acid + carbonate $\longrightarrow$ carbon dioxide + water + salt
   e.g. $2HCl + FeCO_3 \longrightarrow CO_2 + H_2O + FeCl_2$

Note: these can be carried out with either strong or weak acids or bases, the only difference being their reaction rates. A strong acid like nitric acid will react more quickly as all the hydronium ions are immediately free to react with a base, metal or carbonate. A weak acid like ethanoic acid will react more slowly but still go towards a complete reaction, as all the hydronium ions formed are being removed from its equilibrium with water forcing the equilibrium towards completion. This takes time, so weak acids will react slower with a base, metal or a carbonate.

You will not need to remember which acids or bases are weak as they will always tell you this in an exam.

ISBN: 9780170260107

EXAMINER'S TIP

You must know how to determine acid-base conjugate pairs in any acid-base reaction.

Example: $H_2SO_4 + H_2O \longrightarrow HSO_4^- + H_3O^+$

H_2SO_4/HSO_4^- Conj. Acid/Conj. Base is one pair; the other is H_2O/H_3O^+ Conj. Base/Conj. Acid. All that differs between the species in each pair is a H^+.

What do the following key words or phrases mean?

Bronsted-Lowry acid	
Bronsted-Lowry base	
Conjugate base	
Conjugate acid	
Hydronium ion	
Neutralisation	
Dissociate	
Ionise	
Proton	
Strong acid	
Weak acid	
Dilute acid	
Concentrated acid	

 ISBN: 9780170260107

3.7: Writing acid-base reactions

1 Complete the following reactions of acids reacting with water.

a $HCl + H_2O \longrightarrow$

b $HNO_3 + H_2O \longrightarrow$

c $H_2SO_4 + H_2O \longrightarrow$

(You might notice there is more than one answer here!)

d $CH_3CH_2COOH + H_2O \rightleftharpoons$

e $HF + H_2O \rightleftharpoons$

f $CH_3COOH + H_2O \rightleftharpoons$

2 Complete the following reactions of bases reacting with water.

a $NaOH_{(aq)} \longrightarrow$

b $Mg(OH)_{2\,(aq)} \longrightarrow$

c $NH_3 + H_2O \rightleftharpoons$

d $CH_3COO^- + H_2O \rightleftharpoons$

3 Determine all the conjugate acid base pairs in 1 and 2 above.

4 Complete the following reactions of acids reacting with either bases, metals or carbonates.

a ____ HCl + ____ $Al(OH)_3 \longrightarrow$

b ____ HCl + ____ $Al_2(CO_3)_3 \longrightarrow$

c ____ HNO_3 + ____ $Zn \longrightarrow$

d ____ CH_3COOH + ____ $PbCO_3 \longrightarrow$

e ____ CH_3COOH + ____ $Mg \longrightarrow$

In 1923 Johannes Nicolaus Brønsted and Thomas Martin Lowry came up with the idea for acids being proton acceptors and bases being proton donors independently at the same time. Both have been credited with the discovery, however often only Brønsted is recognised in textbooks as his name comes first in the alphabet and chemists like to shorten names!

EXPERIMENT 4

The strengths of acids and their reactions

AIM: To look at the differing properties between a strong and a weak acid.

EQUIPMENT and CHEMICALS:

1 $molL^{-1}$ hydrochloric acid
1 cm strips of magnesium
splint
test tube rack
1 $molL^{-1}$ ethanoic acid
marble chips
bung and delivery tube
boiling tubes
Universal indicator
limewater
test tubes

SAFETY PRECAUTIONS:

Hydrochloric acid is corrosive. Universal indicator is flammable.

METHOD and RESULTS:

1 Colours in Universal indicator

Add a few drops of Universal indicator into two test tubes. Add 1 mL of hydrochloric acid to one of the test tubes and 1 mL of ethanoic acid to the other.

Acid	Colour in Universal indicator	pH (approximately)
Hydrochloric acid		
Ethanoic acid		

2 Reaction with magnesium

Place one piece of magnesium into each of two test tubes. To one test tube add 2 mL of hydrochloric acid. Cover with your thumb and then add a lit splint when you feel the pressure build up (the 'pop' test for hydrogen gas). Now into the other test tube add 2 mL of ethanoic acid and repeat the pop test.

Acid	Observations with magnesium
Hydrochloric acid	
Ethanoic acid	

3 Reaction with calcium carbonate (marble)

Place one lump of marble into each of two boiling tubes. To one boiling tube add 2 mL of hydrochloric acid, seal with a bung and place the delivery tube connected to the bung into 3 – 4 mL of limewater (the carbon dioxide gas test). Now repeat with the other boiling tube and 2 mL of ethanoic acid.

ISBN: 9780170260107

Acid	Observation with calcium carbonate
Hydrochloric acid	
Ethanoic acid	

QUESTIONS:

1 Which acid is the strong acid and which is the weak acid? How do you know?

2 Complete the equations below for the reactions above:

a $HCl + H_2O \longrightarrow$

b $CH_3COOH + H_2O \rightleftharpoons$

c ___ HCl + ___ $Mg \longrightarrow$

d ___ CH_3COOH + ___ $Mg \longrightarrow$

e ___ HCl + ___ $CaCO_3 \longrightarrow$

f ___ CH_3COOH + ___ $CaCO_3 \longrightarrow$

3 Predict what would happen if you used a dilute and a concentrated version of hydrochloric acid to complete the reactions above. What would you observe and why?

4 Predict what would happen if you used a dilute and a concentrated version of ethanoic acid to complete the reactions above. What would you observe and why?

ISBN: 9780170260107

CHEMISTRY APPS

Some common acids and where you might find them:

Acid	Plant origin	Acid	Animal origin
Citric acid	Orange and lemon	Hydrochloric acid	Human stomach
Tartaric acid	Grapes	Lactic acid	Muscles during exercise
Ascorbic acid	Vitamin C in citrus fruits and blackcurrants	Uric acid	Urine
Methanoic acid	Nettle sting	Methanoic acid	Ant

What is pH?

pH is a measure of the concentration of hydronium ions in a solution, calculated in $molL^{-1}$. By looking at a pH value we can tell whether a substance is strongly acidic (pH 1 – 3), weakly acidic (pH 4 – 6), neutral (which means neither acidic nor basic so the concentrations of hydronium ions must be equal to the concentration of hydroxide ions) (pH 7), weakly basic (pH 8 – 10) or strongly basic (pH 11 – 14). The pH scale below shows some common substances and their pH values.

Concentration of hydronium ions compared to distilled water		pH	Examples of solutions and their respective pH
	1/10,000,000	14	drain cleaner
	1/1,000,000	13	bleach, oven cleaner
	1/100,000	12	soapy water
	1/10,000	11	ammonia solution
	1/1,000	10	
	1/100	9	baking soda
	1/10	8	sea water
	0	7	pure water
	10	6	urine, saliva
	100	5	black coffee
	1000	4	acid rain
	10,000	3	lemon juice, vinegar, orange juice
	100,000	2	
	1,000,000	1	HCl in stomach acid
	10,000,000	0	battery acid

ISBN: 9780170260107

Acid calculations

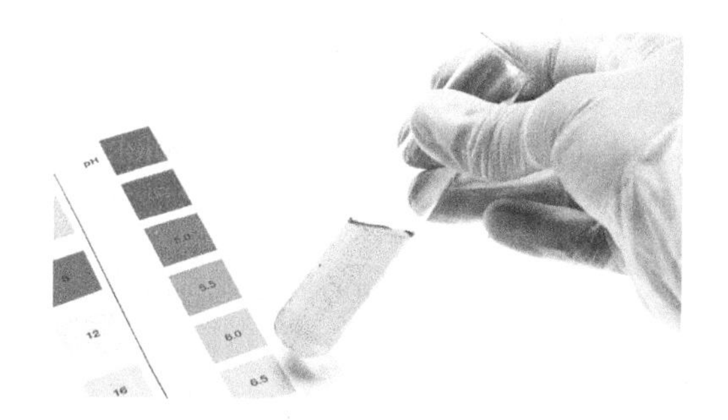

We can calculate the pH of an acid by using a simple formula.
You will need to remember this formula for the exam:

$$pH = -\log[H_3O^+]$$

If it is a strong acid all of the available H^+ of the original acid will become H_3O^+, since strong acids completely dissociate. You will not need to calculate the pH of a weak acid at NCEA Level 2.

Example:
If you have 0.1 $molL^{-1}$ hydrochloric acid, what will be the pH?

$pH = -\log[0.1] = 1$ **Remember pH does not have units.**

We can also calculate the concentration of hydronium ions and therefore the concentration of acid used if we know the pH, using this formula:

$[H_3O^+]$ = inverse log – pH *(To get inverse log use the shift button in your calculator and it should look like this 10^.)*

Example:
If you have a solution of hydrochloric acid with a pH of 3, what is the concentration of hydronium ions?

$[H_3O^+]$ = inverse log - 3 = 0.001 $molL^{-1}$

3.8: Acid calculations

1 Calculate the pH of the following solutions of acid:

a 1 $molL^{-1}$ hydrochloric acid solution

b 0.01 $molL^{-1}$ hydrochloric acid solution

c 0.0122 $molL^{-1}$ hydrochloric acid solution

d 1.45 $molL^{-1}$ nitric acid solution

e 2.13 $molL^{-1}$ nitric acid solution

f 0.478 $molL^{-1}$ sulfuric acid solution (Take care!)

2 Calculate the concentration of hydronium ions in $molL^{-1}$ for the following solutions:

a pH 1 nitric acid solution

b pH 4.55 nitric acid solution

ISBN: 9780170260107

c pH 3.26 hydrochloric acid solution

d pH 6.66 hydrochloric acid solution

e pH 0.554 sulfuric acid solution

f pH 1.27 sulfuric acid solution

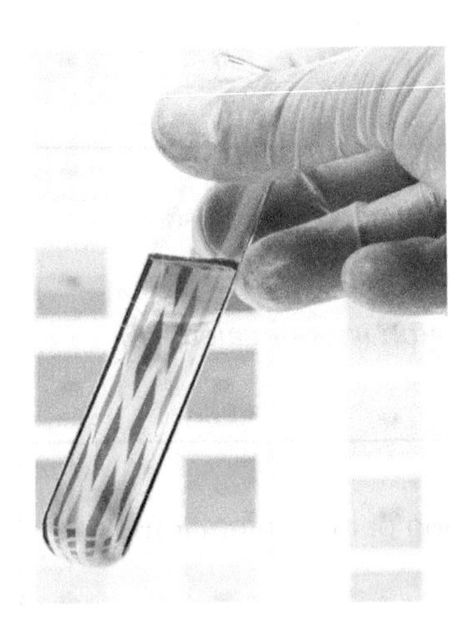

Base pH calculations

Calculating the pH of a base requires *two steps* as in order to calculate the pH we still need to know the concentration of hydronium ions.

Step 1: Find the concentration of hydronium ions.

In pure water there are equal concentrations of both H_3O^+ and OH^- and in fact water is in an equilibrium state.

$$2H_2O \rightleftharpoons OH^- + H_3O^+$$

The equilibrium constant for this K_w is always equal to 1×10^{-14}, so using this we can calculate the concentration of hydronium ions using this constant.

$$K_w = [H_3O^+][OH^-] = 1 \times 10^{-14} \quad \text{or} \quad [H_3O^+] = \frac{1 \times 10^{-14}}{[OH^-]}$$

Step 2: Calculate the pH.

Now that we know the concentration of hydronium ions we can just use the same equation as we did for calculating the pH of an acid.

$$pH = -\log [H_3O^+]$$

> *Example:*
> What is the pH of a 0.100 $molL^{-1}$ sodium hydroxide solution?
>
> **Step 1:** $[H_3O^+] = K_w / [OH^-] = (1 \times 10^{-14}) / 0.100 = 1 \times 10^{-13}$
>
> **Step 2:** $pH = -\log[H_3O^+] = 13$

We can also calculate the $[OH^-]$ in the original base solution by reversing the process above.

Step 1: Calculate the $[H_3O^+]$

$$[H_3O^+] = \text{inverse log} - pH$$

Step 2: Calculate the $[OH^-]$

$$[OH^-] = Kw / [H_3O^+]$$

> *Example:*
> What is the concentration of sodium hydroxide if the pH of the solution is 13.9?
>
> **Step 1:** $[H_3O^+] = \text{inverse log} - pH = \text{inverse log} - 13.9$
>
> **Step 2:** $[OH^-] = K_w / [H_3O^+] = 1 \times 10^{-14} / 1.26 \times 10^{-14} = 0.794 \text{ mol L}^{-1}$

ISBN: 9780170260107

What do the following key words or phrases mean?

pH	
K_w	

3.9: Base calculations

1 Calculate the pH of the following basic solutions.

a 0.100 $molL^{-1}$ sodium hydroxide solution

b 0.0160 $molL^{-1}$ sodium hydroxide

c 1.20 $molL^{-1}$ sodium hydroxide

d 0.0126 $molL^{-1}$ zinc hydroxide

e 2.00 $molL^{-1}$ aluminium hydroxide

2 Calculate the concentration in $molL^{-1}$ of base in the following solutions.

a sodium hydroxide pH 8.66

b sodium hydroxide pH 10.1

c sodium hydroxide pH 14.0

d copper hydroxide pH 11.6

e zinc hydroxide pH 13.8

Acid base explanations

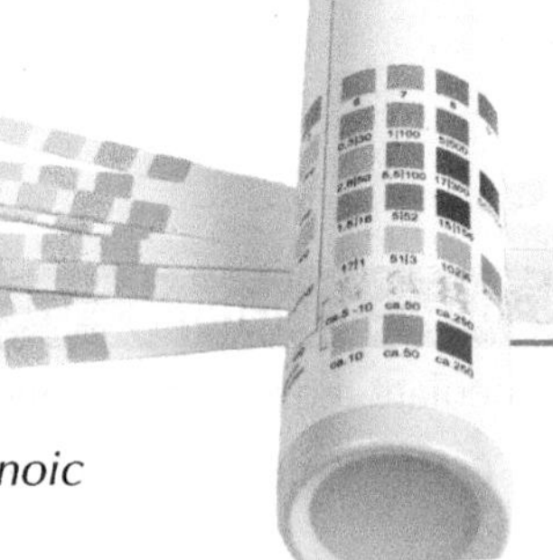

You will need to explain (by looking at pH values and or concentrations of acids and bases) whether they are strong or weak and so what you might observe when they react and how conductive they might be. *Remember weak acids will react more slowly than strong ones and will have pH values higher than 3. They will also be less conductive as there will be fewer ions available in solution to react with.*

Example:

Explain the difference between 1 $molL^{-1}$ solutions of hydrochloric acid and ethanoic acid and their reactions with sodium carbonate.

$$HCl + H_2O \longrightarrow Cl^- + H_3O^+$$

$$2HCl + Na_2CO_3 \longrightarrow 2NaCl + CO_2 + H_2O$$

The hydrochloric acid reaction shown above will have more vigorous fizzing and will have a faster rate of reaction, as hydrochloric acid is a strong acid which will completely dissociate in water to produce hydronium and chloride ions. This means all the hydronium ions are available to react with the base at the start of the reaction.

$$CH_3COOH + H_2O \rightleftharpoons CH_3COO^- + H_3O^+$$

$$2CH_3COOH + Na_2CO_3 \longrightarrow 2NaCH_3COO + CO_2 + H_2O$$

On the other hand, ethanoic acid is a weak acid and so will react much more slowly with less vigorous bubbling despite having the same concentration. This is because ethanoic acid only partially dissociates in water so not all the hydronium ions are available to react with the sodium carbonate at the beginning. This causes a slower reaction which will eventually go to completion once it has forced the equilibrium between the ethanoic acid and water towards the products by removing all the H_3O^+ formed.

3.10: Acid base explanations

1 Explain the reaction between 1 $molL^{-1}$ solutions of nitric acid and ethanoic acid and magnesium metal.

 ISBN: 9780170260107

2 Explain the difference between the strength and conductivity of 0.1 $molL^{-1}$ solutions of propanoic acid and sulfuric acid. (Propanic acid is a weak acid.)

CHECKPOINT 3
WHAT HAVE YOU LEARNED SO FAR?

Acids and bases

1 Compare and contrast the words *concentration* and *strength*.

2 Below are the pH values of some common acids and bases. Write whether each is strong or weak in the space provided. Then write the equation for each one's (except milk and blood) reaction with water.

	Acid/Base	pH	Weak/Strong
A	Hydrochloric acid	1	
B	Propanoic acid	3	
C	Milk	6	
D	Blood	8	
E	Lime water/Calcium hydroxide	10	
F	Sodium hydroxide	14	

ISBN: 9780170260107

CHECKPOINT 3

WHAT HAVE YOU LEARNED SO FAR?

3 Complete the equations below:

a $CH_3COOH + NaOH \longrightarrow$ ____________________

b $H_2SO_4 + KOH \longrightarrow$ ____________________

4 Name three common uses for acids and bases.

____

____

____

5 In the boxes below draw all the species (molecules and ions) present and their relative amounts when HCl is dissociated in water and when NH_3 is dissociated in water.

$HCl_{(aq)}$ $NH_{3(aq)}$

FACT

In 1909 the Danish scientist Soren Sörensen devised the pH scale, the one we know and still use today. Previously acid measurements were based around how much acid was used, rather than the concentration of hydrogen ions in solution. Sörensen realised that some acids dissociate more than others. The letter 'p' in pH originally was short for *potenz* (German) or perhaps for *puissance* (French), both of which mean *power*. H of course is hydrogen.

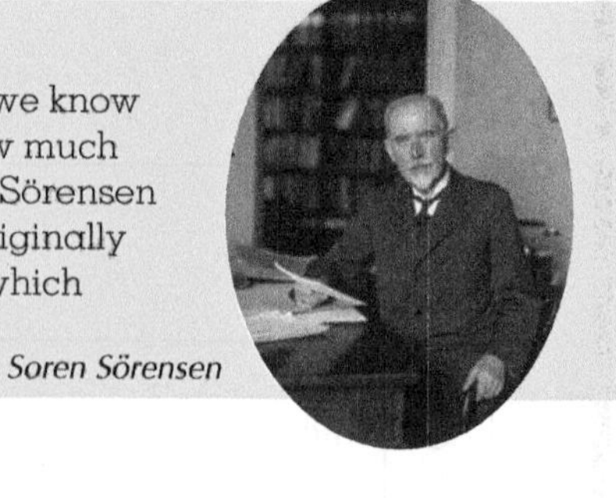

Soren Sörensen

 ISBN: 9780170260107

EXAM-TYPE QUESTIONS

Before you get started here are some points to remember from the Examiner's Report:

- Don't use generalisations, use terms carefully.
- Plan your answer before you start.

Question one

The Contact Process is an important industrial process used for making sulfuric acid. One of the steps for this process is shown below.

$$2SO_{2(g)} + O_{2(g)} \rightleftharpoons 2SO_{3(g)}$$

Explain two ways in which the manufacturers could increase the rate of reaction. Use collision theory to illustrate your answer.

Question two

The iodine clock reaction involves two colourless solutions turning black about a minute after being mixed. A magician wants to use this in his upcoming show and get an audience member to come up and say the magic word for the transformation to occur. The equation for the clock reaction is:

$$2\ I^-_{(aq)} + S_2O_8^{2-}{}_{(aq)} \longrightarrow I_{2\ (aq)} + 2\ SO_4^{2-}{}_{(aq)}$$

Explain two ways in which the rate of reaction can be slowed down by the magician in order to allow time for the audience member to come up to the stage and say the word.

Question three

The following results were obtained from a rates of reaction investigation. The equation below shows the reaction that was carried out:

$$H_2SO_{4\,(aq)} + Zn_{(s)} \longrightarrow ZnSO_{4\,(aq)} + H_{2\,(g)}$$

	Experiment 1: time taken (s) for the zinc to disappear with $c(H_2SO_4) = 2\ molL^{-1}$	Experiment 2: time taken (s) for the zinc to disappear with $c(H_2SO_4) = 1\ molL^{-1}$
Trial 1	10	21
Trial 2	11	22
Trial 3	10	21
Average	**10.3**	**21.3**

a Explain what was being investigated and what the results indicate. Use your knowledge and understanding of rates of reaction and collision theory.

b Explain one other way you could increase the rate of this reaction other than the one you explained for question **3a**. Use your knowledge and understanding of rates of reaction and collision theory to illustrate your answer.

ISBN: 9780170260107

EXAM-TYPE QUESTIONS

Question four

The decomposition of calcium carbonate into calcium oxide and carbon dioxide gas is shown below as an example of a dynamic equilibrium.

$$CaCO_{3\,(s)} \rightleftharpoons CO_{2\,(g)} + CaO_{(s)} \qquad \Delta H = +ve\ kJmol^{-1}$$

a Write the K_c expression for this equilibrium.

b Discuss what would happen to the equilibrium above if it was heated.

c Explain what would happen if a catalyst was added to the value of K_c.

ISBN: 9780170260107

Question five

Nitrogen dioxide, NO_2 is a brown gas and nitrogen tetraoxide, N_2O_4 is colourless.

$$2NO_{2(g)} \rightleftharpoons N_2O_{4(g)} \qquad \Delta H = -54 \text{ kJ mol}^{-1}$$

a Write the equilibrium constant expression, K_c, for this reaction.

b Calculate the value of K_c if there is a 0.100 $molL^{-1}$ concentration for both the nitrogen dioxide and the nitrogen tetroxide.

c Is the above K_c value reactant or product favoured? How do you know?

d Predict what you would observe if some NO_2 was removed from the above equilibrium system. Explain.

e Predict what you would observe if the equilibrium was carried out at a lower temperature. Explain.

 ISBN: 9780170260107

EXAM-TYPE QUESTIONS

Question six

Aqueous iron (III) ions can form a complex ion with chloride ions to form the tetrachloroferrate(III) ion. The equilibrium for this is shown below with its K_c value:

$$Fe^{3+}_{(aq)} + 4Cl^{-}_{(aq)} \rightleftharpoons FeCl_{4\,(aq)}^{-} \qquad K_c = 8.00 \times 10^{-2}$$

a Write the K_c expression for this equilibrium.

b If the concentration of the chloride ion is 0.801 mol L^{-1} and that of the iron(III) ion 0.220 mol L^{-1}, calculate the concentration of the tetrachloroferrate(III) complex ion.

c Explain what would happen to this equilibrium system if the concentration of chloride ions was halved.

Question seven

Iodine dissolves in a solution of potassium iodide to form the I_3^- ion, shown in the equilibrium below with its K_c value:

$$I^{-}_{(aq)} + I_{2(aq)} \rightleftharpoons I_{3\,(aq)}^{-} \qquad K_c = 7.11 \times 10^{-2}$$

a Write the equilibrium constant expression K_c for this equilibrium.

b Given the value of K_c shown above, circle the species below that would be in higher concentration for this particular K_c value. Explain why you chose this answer.

I^- $\qquad$ I_3^-

Explanation:

ISBN: 9780170260107

c If the concentration of the I^- ion is 0.125 mol L^{-1}, and that of the I_3^- ion is 0.156 mol L^{-1}, calculate the concentration of iodine.

Question eight

pH (Acid X) = 1.02	pH (Acid Y) = 5.03

Acid X and Acid Y have the pH values given above. Assume that the solutions of X and Y have the same concentration. Describe and compare the reaction of equal amounts of these two acid solutions with calcium carbonate.

In your discussion comment on the initial pH of each solution and also comment on the relative amount of calcium carbonate used up in each reaction.

 ISBN: 9780170260107

EXAM-TYPE QUESTIONS

Question nine

For the following reactions, label the conjugate acid base pairs.

a $HCl_{(g)} + H_2O_{(l)} \longrightarrow H_3O^+_{(aq)} + Cl^-_{(aq)}$

b $NaOH_{(aq)} + HCl_{(aq)} \longrightarrow NaCl_{(aq)} + H_2O_{(l)}$

c $CH_3COOH_{(l)} + H_2O_{(l)} \rightleftharpoons CH_3COO^-_{(aq)} + H_3O^+_{(aq)}$

d Explain why ethanoic acid is in equilibrium with water, whereas the reaction between hydrochloric acid with water goes to completion.

e Explain why sodium hydroxide is neutralised more quickly by hydrochloric acid than by ethanoic acid.

ISBN: 9780170260107

Question ten

Explain the difference between the strength and the conductivity of the following 1 $molL^{-1}$ solutions. Refer to the pH values shown below.

Hydrochloric acid pH	Ethanoic acid pH
1.00	3.00

ISBN: 9780170260107

INTERNAL STANDARDS REVIEW

2.1 – Carry out quantitative analysis (91161)

There are *two* parts to completing this internal standard successfully – a **titration** where you must accurately determine the concentration of an unknown solution and a set of **calculations**.

Here are the main things you need to know in order to get the grade you desire:

Achieved

In the titration: You must get three titre volumes within 0.4 mL of each other, that is within 0.8 mL of the expected outcome.

Completing a titration:

Equipment required: pipette, pipette filler, burette, conical flask, bottle of distilled water, indicator, acid, base, funnel (optional to fill the burette).

1 Clean all equipment with **distilled** water before you begin and clean the **burette** with the solution that is going to be poured into it and the **pipette** with the solution that will be used in it.
2 Pipette a given quantity (this is called an **aliquot**) of either an acid or a base into a clean, dry conical flask and add a few drops of **indicator** (usually phenolphthalein or methyl orange will be used).
3 Fill the burette and record the volume it reaches to two decimal places with either a 0.05 or 0.00 on the end.
4 Carefully pour some of the contents of the burette into the conical flask until the indicator changes colour in one drop.
5 Record the final volume it reaches. The initial volume subtracted from the final volume is your **titre volume**.
6 Repeat until you get three results that are within **0.4 mL** of each other.

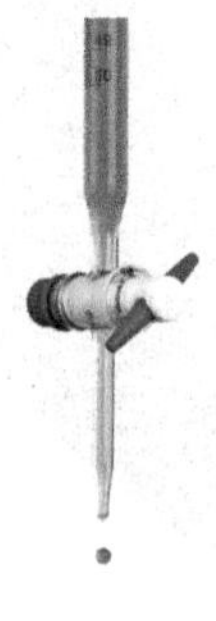

Burette

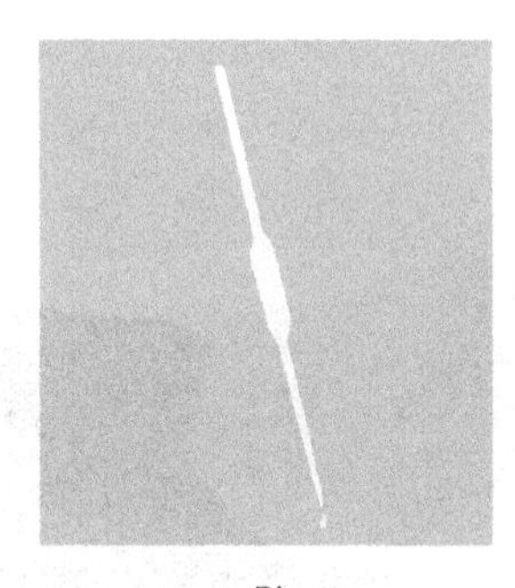

Pipette

Conical flask

In the calculations: You must be able to use and rearrange the following two equations in order to calculate a given quantity.

Moles (n) = mass(m)/molar mass(M)

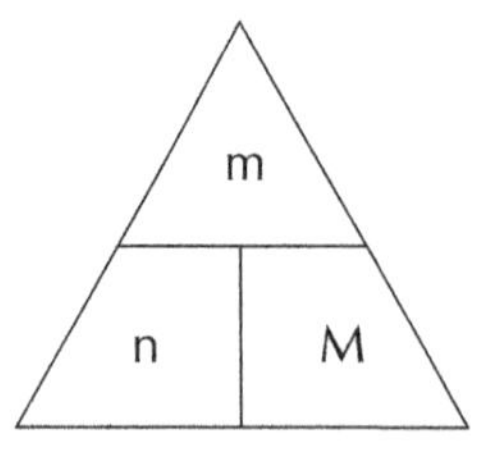

concentration (c) = moles (n)/volume(V)

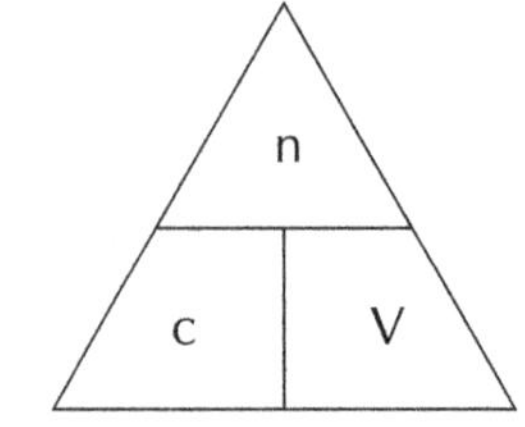

ISBN: 9780170260107

Achieved	**Example calculation:** *What is the mass of 5.00 mol of NaOH?* M(NaOH) = 23.0 + 16.0 + 1.00 = 40.0 $gmol^{-1}$ *(using their atomic masses from the periodic table)* m(NaOH) = n(NaOH) x M(NaOH) = 5.00 x 40.0 = 200 g
Merit	**In the titration:** You must get 3 titre volumes within 0.4 mL of each other, that is within 0.5 mL of the expected outcome. **In the calculations:** You must complete a two step calculation. **Example calculation:** *Calculate the concentration of the new solution of HCl when 700 mL of 0.100 $molL^{-1}$ of HCl is mixed with 250 mL of 0.200 $molL^{-1}$ HCl.* **Step 1: Calculate the total volume** – (700 + 250)/1000 = 0.950 L *(make sure you convert every volume into Litres).* **Step 2: Calculate the total number of moles – c_1 x V_1 + c_2 x V_2 = Total number of moles** n_{Total} = 0.700 x 0.100 + 0.250 x 0.200 = 0.120 mol $c_{Total} = n_T/V_T$ = 0.120/0.950 = 0.126 $molL^{-1}$ (3 s.f.)
Excellence	**In the titration:** You must get 3 titre volumes within 0.2 mL of each other, that is within 0.2 mL of the expected outcome. **In the calculations:** You must complete a calculation with more than two steps to it and write your answer to 3 significant figures and with correct units. **Example calculation:** Calculate the concentration of HCl when 20.0 mL of it was titrated with the following titre volumes of 0.100 $molL^{-1}$ Na_2CO_3 20.20, 20.30, 20.95 and 20.35mL. $Na_2CO_3 + 2HCl \longrightarrow 2NaCl + H_2O + CO_2$ **Step 1: Calculate the average titre volume of Na_2CO_3** *(ignore any outliers in this case 20.95mL)* (20.20 + 20.30 + 20.35)/3 = 20.2833 mL *(don't round until the end)* **Step 2: Calculate the moles of Na_2CO_3 used** (Remember n = c x V) $n(Na_2CO_3)$ = 0.100 x 20.2833/1000 = 0.0020283 mol **Step 3: Using the mole ratio from the equation calculate the moles of HCl** $n(Na_2CO_3)$: n(HCl) 1 : 2 therefore the moles of HCl = 2 x $n(Na_2CO_3)$ = 0.0040566 mol **Step 4: Calculate the concentration of HCl (remember c = n/V)** c(HCl) = n(HCl)/V(HCl) = 0.0040566/(20/1000) = 0.203 $molL^{-1}$ (3 s.f.)

ISBN: 9780170260107

2.2 – Carry out procedures to identify ions in solution (91162)

Here are the main things you need to know in order to get the grade you desire.

Achieved	Identify the ions present from a given unknown solution, by writing the correct observations you have seen with the **name or the formula of every precipitate** that is formed. **Example answer:** *Identifying the iron (III) ion, Fe^{3+}* An orange precipitate of $Fe(OH)_3$ or iron (III) hydroxide forms when 2 drops of NaOH are added, which remains in excess NaOH. When a new sample of the ion is added to KSCN a blood-red solution forms.
Merit	Identify the ions present from a given unknown solution, by writing the correct observations you have seen with the **correctly balanced symbol equation** of the precipitate formed. **Example answer:** *Identifying the iron (III) ion, Fe^{3+}* An orange precipitate of $Fe(OH)_3$ (iron (III) hydroxide) forms when 2 drops of NaOH are added, which remains in excess NaOH. $Fe^{3+}_{(aq)} + 3OH^-_{(aq)} \longrightarrow Fe(OH)_{3\ (s)}$ When a new sample of the ion is added to KSCN a blood red-solution forms.
Excellence	Give a comprehensive justification of which ions are present in a given unknown solution by linking to observations and **balanced** observations and **balanced equations for both the precipitates and the complex ions formed.** **Example answer:** *Identifying the iron (III) ion, Fe^{3+}* An orange precipitate of $Fe(OH)_3$ or iron (III) hydroxide forms when 2 drops of NaOH are added, which remains in excess NaOH. $Fe^{3+}_{(aq)} + 3OH^-_{(aq)} \longrightarrow Fe(OH)_{3\ (s)}$ When a new sample of the ion is added to KSCN a blood-red solution forms. $Fe^{3+}_{(aq)} + SCN^-_{(aq)} \longrightarrow [FeSCN]^{2+}_{(aq)}$

  ISBN: 9780170260107

Complex ion equations

There is no easy way to learn these other than to memorise them. To help, make up posters of this list and stick them all over your bedroom wall to help remind you constantly what the equations look like. Use colour to help you remember each individual complex ion. You don't need to be able to name the complex ions.

Name	Equation	Observation
zincate ion	$Zn(OH)_{2(s)} + 2OH^-_{(aq)} \longrightarrow \mathbf{[Zn(OH)_4]^{2-}_{(aq)}}$ OR $Zn^{2+}_{(aq)} + 4OH^-_{(aq)} \longrightarrow \mathbf{[Zn(OH)_4]^{2-}_{(aq)}}$	colourless solutions
tetrammine zinc (II) ion	$Zn(OH)_{2(s)} + 4NH_{3(aq)} \longrightarrow \mathbf{[Zn(NH_3)_4]^{2+}_{(aq)}} + 2OH^-_{(aq)}$ OR $Zn^{2+}_{(aq)} + 4NH_{3(aq)} \longrightarrow \mathbf{[Zn(NH_3)_4]^{2+}_{(aq)}}$	
plumbate ion	$Pb(OH)_{2(s)} + 2OH^-_{(aq)} \longrightarrow \mathbf{[Pb(OH)_4]^{2-}_{(aq)}}$ OR $Pb^{2+}_{(aq)} + 4OH^-_{(aq)} \longrightarrow \mathbf{[Pb(OH)_4]^{2-}_{(aq)}}$	
aluminate ion	$Al(OH)_{3(s)} + OH^-_{(aq)} \longrightarrow \mathbf{[Al(OH)_4]^-_{(aq)}}$ OR $Al^{3+}_{(aq)} + 4OH^-_{(aq)} \longrightarrow \mathbf{[Al(OH)_4]^-_{(aq)}}$	
diammine silver (I) ion	$AgCl_{(s)} + 2NH_{3(aq)} \longrightarrow \mathbf{[Ag(NH_3)_2]^+_{(aq)}} + Cl^-_{(aq)}$	
tetrammine copper (II) ion	$Cu^{2+}_{(aq)} + 4NH_{3(aq)} \longrightarrow \mathbf{[Cu(NH_3)_4]^{2+}_{(aq)}}$ OR $Cu(OH)_{2(s)} + 4NH_{3(aq)} \longrightarrow \mathbf{[Cu(NH_3)_4]^{2+}} + 2OH^-_{(aq)}$	deep blue solution
iron thiocyanate ion	$Fe^{3+}_{(aq)} + SCN^-_{(aq)} \rightleftharpoons \mathbf{[FeSCN]^{2+}_{(aq)}}$	deep red solution

ISBN: 9780170260107

2.3 – Demonstrate understanding of the chemistry used in the development of a current technology (91163)

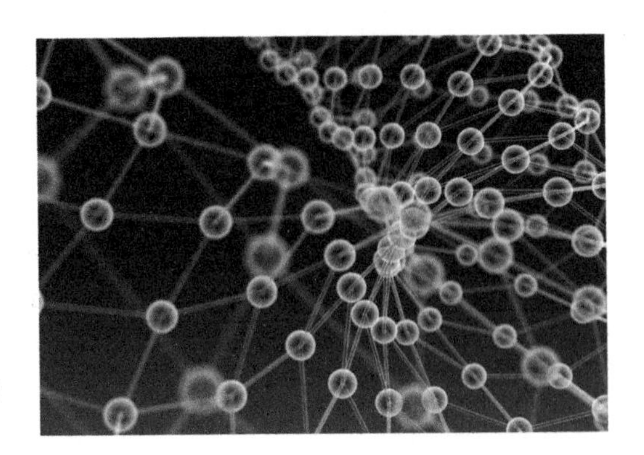

Current technologies (that is technologies we use today) that you could study include: Conducting polymers, nanotechnology, cosmetics, pharmaceuticals, paints, polymers, catalytic converters, fabric and fibre technology and alloys.

Here are the main things you need to know in order to get the grade you desire:

Achieved	Process and interpret information to provide an account of the chemistry used in the development of a current technology using chemistry vocabulary, symbols and conventions. **For example:** Your report provides only a **brief outline of the main points related to the chemistry** of that technology and how it was developed; uses some chemistry vocabulary to describe these points. AND, you have shown to your teacher that you have **read and processed** some information on the subject.
Merit	Make and explain the links between the technology and the chemistry of it using chemistry vocabulary, symbols and conventions. **For example:** Your report for merit needs **more depth** with more relationships shown between the current technology and chemistry knowledge. AND, you should have **thoroughly processed** your information on the subject.
Excellence	Evaluate how the development of that technology was made due to the knowledge of chemistry the inventor had. **For example:** The main difference between merit and excellence reports are that your report covers **all the related chemistry** to the development of the technology with explanations of how they link and relate. AND, you **evaluate** the usefulness of that technology to society.

ISBN: 9780170260107

2.7 – Demonstrate understanding of oxidation-reduction (91167)

You need to be able to identify which species (by name or correct formula) in a given equation is being reduced and which is being oxidised by linking to either the loss or gain of electrons or the oxidation numbers.

Achieved	*You need to be able to identify which species in a given equation is being reduced and which is being oxidised, you will need to explain with evidence, either from loss or gain of electron oxidation numbers.*

To do this you need oxidation numbers, here are the rules for working out the oxidation number of a species:

1. All elements in their standard states are **zero**. *E.g. Zn or Cl_2.*
2. Hydrogen is always **+1** unless it is part of a metal hydride in which case it becomes **-1**. *E.g. the H in H_2O is +1, whereas the H is NaH is -1.*
3. Oxygen is always **-2**, unless it is in H_2O_2, (hydrogen peroxide) where it becomes **-1**. *E.g. the O in H_2O is -2, whereas the O in H_2O_2 is -1.*
4. The sum of the oxidation numbers in a neutral compound must equal **zero**. *E.g. in H_2O there are 2Hs each with a +1 state and there is an O with a -2 state.*
5. In compounds the **normal charge** on ions is their oxidation numbers. *E.g. the Na in NaCl is +1 and the Cl is -1.*
6. In a polyatomic ion the sum of the oxidation numbers must equal the **charge on that ion**. *E.g. in MnO_4^-, each O has a -2 oxidation number, there are four of them in total, meaning there is a -8 charge from the oxygen, therefore the Mn must be +7 if the total oxidation number for the ion is to equal -1.*

Oxidation is the INCREASE in oxidation number. Reduction is the DECREASE in oxidation number.

For example:

In the following reaction, which species is oxidised and which is reduced, and what observations would be made when it occurs?

$$MnO_4^- + Fe^{2+} \longrightarrow Mn^{2+} + Fe^{3+}$$

The MnO_4^- is a purple solution and it is reduced to Mn^{2+} which is a colourless ion.

The Fe^{2+} ion is green and is oxidised to Fe^{3+} which is yellow.

ISBN: 9780170260107

Merit	

Merit

You need to be able to balance half equations as well as link the observations and the oxidation numbers to the oxidation and reduction half equations.

Balancing redox equations:

1. Separate into the two half equations and balance atoms other than O or H.
   E.g. $Cr_2O_7^{2-} + Fe^{2+} \longrightarrow Cr^{3+} + Fe^{3+}$
   $Cr_2O_7^{2-} \longrightarrow 2Cr^{3+}$ $\quad$ $Fe^{2+} \longrightarrow Fe^{3+}$
2. Balance O atoms by adding H_2O.
   E.g. $Cr_2O_7^{2-} \longrightarrow 2Cr^{3+} + 7H_2O$
3. Balance H atoms by adding H^+.
   E.g. $Cr_2O_7^{2-} + 14H^+ \longrightarrow 2Cr^{3+} + 7H_2O$
4. Balance the charge by adding electrons.
   E.g. $Cr_2O_7^{2-} + 14H^+ + 6e^- \longrightarrow 2Cr^{3+} + 7\,H_2O$ $\quad$ $Fe^{2+} \longrightarrow Fe^{3+} + e^-$
5. Multiply out the electrons and add the two equations. *(Note you may also be able to cancel out H_2O and H^+ atoms if they are on both sides of the equation.)*
   $6Fe^{2+} \longrightarrow 6Fe^{3+} + 6e^-$
   $Cr_2O_7^{2-} + 14H^+ + 6Fe^{2+} \longrightarrow 2Cr^{3+} + 7H_2O + 6Fe^{3+}$

For example:

In the following reaction, which species is oxidised and which is reduced, and what observations would be made when it occurs?

$$MnO_4^- + Fe^{2+} \longrightarrow Mn^{2+} + Fe^{3+}$$

The MnO_4^- is a purple solution and it is reduced from +7 to +2 in Mn^{2+} which is a colourless ion.

$$MnO_4^- + 8H^+ + 5e^- \longrightarrow Mn^{2+} + 4H_2O$$

The Fe^{2+} is green and is oxidised from +2 to +3 in Fe^{3+} which is yellow.

$$Fe^{2+} \longrightarrow Fe^{3+} + e^- \quad (x5)$$

$$MnO_4^- + 8H^+ + 5Fe^{2+} \longrightarrow Mn^{2+} + 4H_2O + 5Fe^{3+}$$

Excellence

You need to link the electron transfer to the observations and whole equations.

For example:

$MnO_4^- + Fe^{2+} \longrightarrow Mn^{2+} + Fe^{3+}$

The MnO_4^- is a purple solution and it is reduced from +7 to +2 in Mn^{2+} which is a colourless ion. Therefore MnO_4^- is the oxidant since it takes electrons away from Fe^{2+}.

$MnO_4^- + 8H^+ + 5e^- \longrightarrow Mn^{2+} + 4H_2O$

The Fe^{2+} is green and is oxidised from +2 to +3 in Fe^{3+} which is yellow. Therefore Fe^{2+} is the reductant since it gives up an electron to MnO_4^-.

$Fe^{2+} \longrightarrow Fe^{3+} + e^- \quad (x5)$

$MnO_4^- + 8H^+ + 5Fe^{2+} \longrightarrow Mn^{2+} + 4H_2O + 5Fe^{3+}$

ISBN: 9780170260107

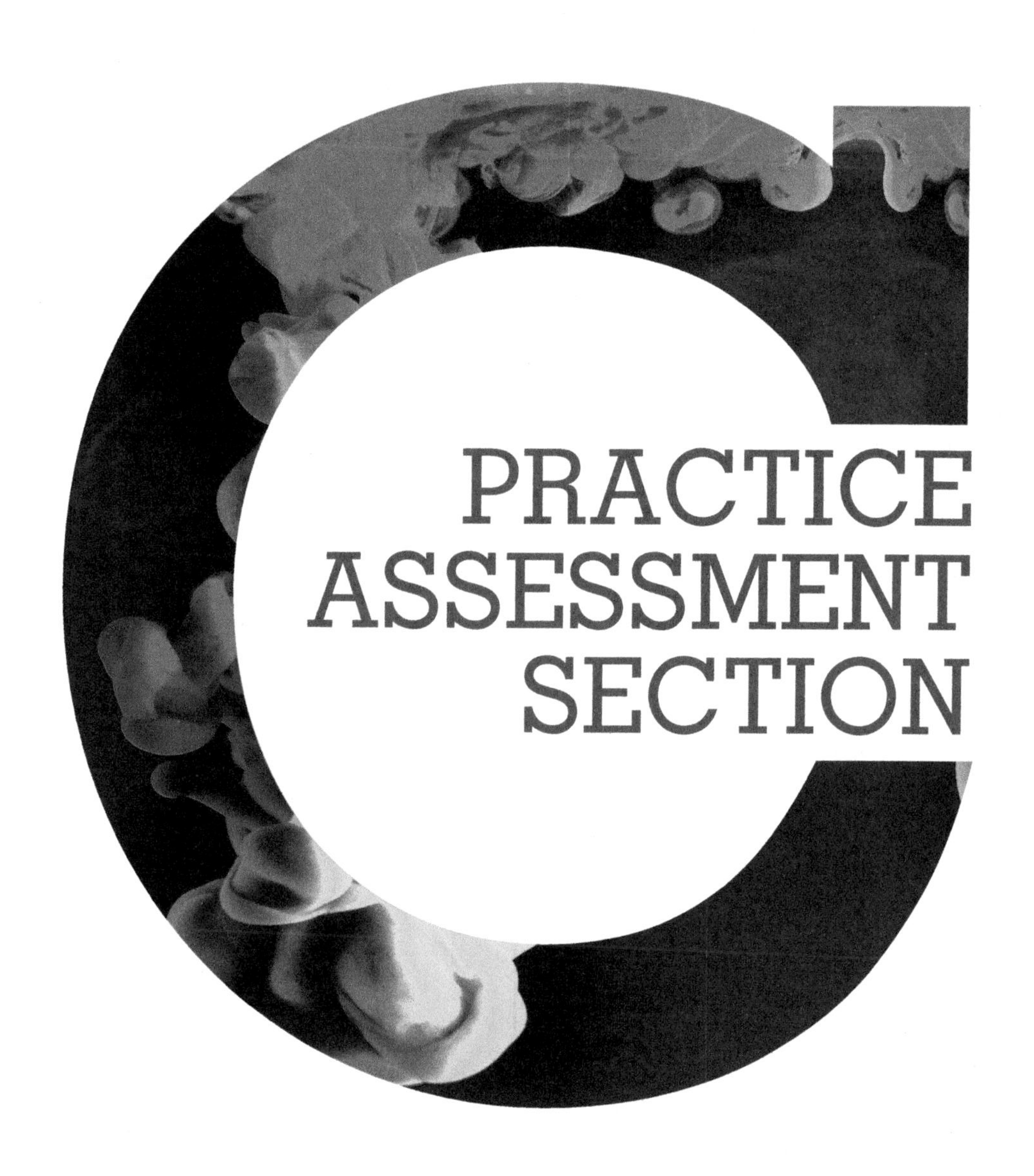

PRACTICE ASSESSMENT SECTION

ISBN: 9780170260107

Level 2 Chemistry

91164 Demonstrate understanding of bonding, structure, properties and energy changes

2

Credits: Five

Achievement	Acheivement with Merit	Achievement with Excellence
Demonstrate understanding of bonding, structure, properties and energy changes.	Demonstrate in-depth understanding of bonding, structure, properties and energy changes.	Demonstrate comprehensive understanding of bonding, structure, properties and energy changes.

You should attempt ALL the questions in this booklet.

You are advised to spend 60 minutes completing this.

TOTAL

QUESTION ONE

TEACHER USE ONLY

a Draw the Lewis diagram (electron dot diagram) and name the shape of the following molecules.

	CH_4	CH_3Cl
Lewis diagram		
Shape		

b Explain why CH_3Cl is a polar molecule, whereas CH_4 is a non-polar molecule even though they have the same shape.

ISBN: 9780170260107

TEACHER USE ONLY

c The following table shows the Lewis structures and bond angles for two molecules BF_3 and NH_3.

	NH_3	BF_3
Lewis diagram	H — N̈ — H, with H bonded below N	:F̤̈: bonded above B; :F̤ and F̤: bonded below B
Bond angle	<109°	120°

Explain why these two molecules have the same number of atoms connected to each central atom and yet they have different bond angles and shapes.

In your answer you should include:

- The shape of NH_3 and BF_3
- Factors which determine the shape of each molecule
- An explanation of why their shapes and bond angles are different despite having the same number of atoms bonded to each central atom.

TEACHER USE ONLY

QUESTION TWO

a Complete the table below by stating the type of particle and bonding between the particles for each substance.

Substance	Type of particle	Bonding
CH_4, methane		
Diamond		
Magnesium		

b Explain in terms of the structure and bonding present in diamond, why diamond is the hardest known substance on earth.

c Explain in terms of the structure and bonding present in methane and magnesium, which one

ISBN: 9780170260107

has the greater melting point and why, and which one can conduct electricity and why.

TEACHER USE ONLY

QUESTION THREE

a State the bonds broken and formed when methane, CH_4 is completely combusted in oxygen.

$CH_{4\ (g)} + 2O_{2\ (g)} \longrightarrow CO_{2\ (g)} + 2H_2O_{\ (l)}$ $\Delta H = -890.3\ kJmol^{-1}$

Bonds broken:

Bonds formed:

b Is the above reaction for the combustion of methane endothermic or exothermic? How do you know?

c If 60 g of methane gas is completely combusted, how much energy would be released?

ISBN: 9780170260107

TEACHER USE ONLY

d Using the bond energy data shown below and the enthalpy value ΔH given above, work out how much energy is required to break a C – H bond.

Bond type	Bond energy ($kJmol^{-1}$)
C – H	?
O = O	495
H – O	467
C = O	799

  ISBN: 9780170260107

Level 2 Chemistry

91165 Demonstrate understanding of the properties of selected organic compounds

Credits: Four

Achievement	Acheivement with Merit	Achievement with Excellence
Demonstrate understanding of the properties of selected organic compounds.	Demonstrate in-depth understanding of the properties of selected organic compounds.	Demonstrate comprehensive understanding of the properties of selected organic compounds.

You should attempt ALL the questions in this booklet.

You are advised to spend 60 minutes completing this.

TOTAL

QUESTION ONE

TEACHER USE ONLY

a Draw the structural isomers of C_4H_9Cl. One of these isomers has been drawn and named for you in the table below. Complete the table to show the structural formulae and IUPAC (systematic) names of the other structural isomers.

H H H H │ │ │ │ H — C — C — C — C — Cl │ │ │ │ H H H H	
1-chlorobutane	

ISBN: 9780170260107

TEACHER USE ONLY

b Using C_4H_9Cl as an example explain what structural isomers are.

c 1- Chlorobutane, $CH_3CH_2CH_2CH_2Cl$, reacts with aqueous KOH and alcoholic NH_3 to form two different products. Compare and contrast the reactions of 1- chlorobutane with the two reagents. In your answer you should include:

- the type of reaction occurring and the reason why it is classified as that type
- the type of functional group formed
- equations showing structural formulae for reactions occurring.

Explain how you could distinguish between the two products formed, using damp litmus.

ISBN: 9780170260107

TEACHER USE ONLY

QUESTION TWO

a Complete the following table to show the structural formula and IUPAC (systematic) name for each compound.

Structural formula	Name
$CH_3-CH_2-CH_2-OH$ (H–C–C–C–O–H, each C bearing two H)	
	Propanoic acid
$CH_3-CHBr-CH_3$ (H–C–C–C–H, Br on the middle C, remaining positions H)	
	Aminomethane

b If you had four unlabelled bottles of each of the four substances on page 171, explain how you could distinguish them using the following reagents – $CaCO_3$ calcium carbonate, $H^+/KMnO_4$ acidified potassium permanganate, damp red litmus paper. In your answer you should include:

- a description of the type of reactions that would occur
- any observations that would be made
- equations showing the structural formulae of the organic reactant(s) and product(s).

TEACHER USE ONLY

ISBN: 9780170260107

QUESTION THREE

TEACHER USE ONLY

a But-2-ene, C_4H_8 has two geometric isomers; draw and name these two isomers in the boxes provided below.

	Isomer 1	**Isomer 2**
Name		
Structural formula		

b Using but-2-ene as an example, explain what geometric isomers are and why they can't be superimposed upon one another.

c If we added HCl to but-1-ene and but-2-ene, explain why two products would form with the but-1-ene but only one would form with the but-2-ene. In your answer you should include:

- equations showing the structural formula and name of the products formed
- an explanation of the type of reaction occurring in each case
- an explanation of which product is the major and why that is according to Markovnikov's rule for but-1-ene.

ISBN: 9780170260107

TEACHER USE ONLY

ISBN: 9780170260107

Level 2 Chemistry

91166 Demonstrate understanding of chemical reactivity

Credits: Four

Achievement	Acheivement with Merit	Achievement with Excellence
Demonstrate understanding of chemical reactivity.	Demonstrate in-depth understanding of chemical reactivity.	Demonstrate comprehensive understanding of chemical reactivity.

You should attempt ALL the questions in this booklet.

You are advised to spend 60 minutes completing this.

TOTAL

QUESTION ONE

TEACHER USE ONLY

The Iodine Clock reaction involves two colourless solutions turning black when added together. The equation for the clock reaction is shown below:

$$2\,I^-_{(aq)} + S_2O_8^{2-}{}_{(aq)} \longrightarrow I_{2\,(aq)} + 2\,SO_4^{2-}{}_{(aq)}$$

a List TWO ways that the rate of this reaction could be ***increased***.

b Explain using your understanding of collision theory how these TWO ways increase the rate of the reaction.

ISBN: 9780170260107

c In another reaction the following results were obtained in the graph below.

$$CaCO_{3\,(s)} + 2\,HCl_{(aq)} \longrightarrow CaCl_{2\,(aq)} + CO_{2\,(g)} + H_2O_{(l)}$$

TEACHER USE ONLY

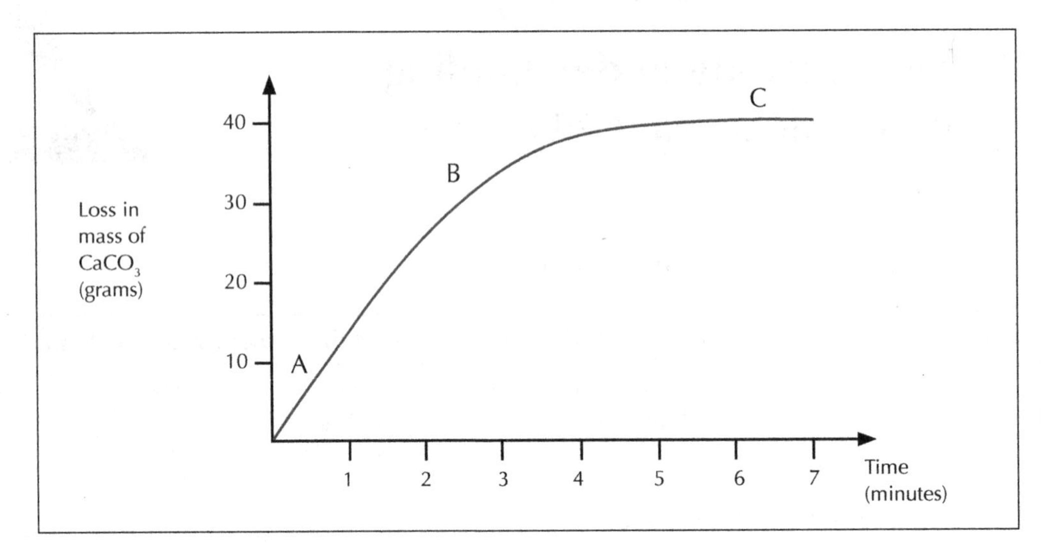

The graph shows the amount of $CaCO_3$ that was lost as the reaction proceeded. Explain why the graph flattens out over time and at which point in the graph there was the fastest rate of reaction: A, B or C.

ISBN: 9780170260107

QUESTION TWO

TEACHER USE ONLY

The Haber process is shown in the equilibrium below:

$$N_{2\,(g)} + 3H_{2\,(g)} \rightleftharpoons 2NH_{3\,(g)}$$

a Write the K_c expression for this reaction.

K_c =

b Explain why a very high pressure is used when this reaction is completed in industry.

c The following table shows how the value of K_c changes for this reaction with temperature.

Temperature (°C)	K_c Value
300	9.6
500	1×10^{-2}

Explain why the K_c value changes with temperature and what this tells us about the concentration of ammonia, NH_3 at equilibrium when this reaction is carried out at these two temperatures.

ISBN: 9780170260107

QUESTION THREE

TEACHER USE ONLY

a Complete the table below of values of hydroxide (OH^-), hydronium (H_3O^+) ions and pH.

	$[OH^-]$ $molL^{-1}$	$[H_3O^+]$ $molL^{-1}$	pH
Solution A	1.02×10^{-6}		
Solution B			13.2

b Justify which solution would conduct electricity better out of solution A and B given the information you calculated above.

c Complete the reactions below of two acids dissociating in water.

Reaction 1	$HCl + H_2O \longrightarrow$	**pH = 1.00**
Reaction 2	$HCO_3^- + H_2O \rightleftharpoons$	**pH = 5.00**

d Justify which acid would react more quickly with magnesium metal given the dissociation reactions and pH values shown above. In your answer you should include:

- an equation of each acid reacting with magnesium
- an explanation for which acid would react faster and why.

ISBN: 9780170260107

TEACHER USE ONLY

GLOSSARY OF TERMS

Addition reaction: an organic reaction where two atoms or small groups of atoms are bonded to the two carbon atoms in a double bond.

Alcohol: group of organic compounds with an OH functional group.

Activation Energy: the energy required for a reaction to occur.

Aliquot: an exact volume of liquid.

Alkane: group of organic molecules made from carbon atoms and hydrogen atoms that are all singly bonded.

Alkene: group of organic compounds made of carbon atoms and hydrogen atoms with one or more double bond(s) between two carbon atoms.

Alkyne: group of organic compounds made of carbon atoms and hydrogen atoms with one triple bond between two carbon atoms.

Amine: group of organic compounds with a NH_2 functional group.

Anion: a negative ion.

Aqueous: when a substance has been dissolved in water.

Asymmetrical: a shape that is not symmetrical *e.g. bent.*

Atom: the smallest neutral unit of matter; with protons, electrons and neutrons.

Atomic number: the number of protons an atom has.

Bent: a molecular shape containing one or two lone pairs, with two bonded pairs of electrons.

Boiling point: the temperature at which a liquid turns into a gas.

Bond: an attraction between two particles – atoms, ions or molecules.

Bond angle: the angle between two bonds.

Bond dipole: a partial charge separation caused between two atoms unequally sharing the electrons in a covalent bond due to an electronegativity difference.

Bond enthalpy: the amount of energy in kJ required to break one mole of a bond in gaseous state.

Bonding pair: pair of electrons that form a covalent bond between two atoms.

Bronsted-Lowry acid: a substance which donates protons.

Bronsted-Lowry base: a substance which accepts protons.

Brittle: a structure that breaks apart easily when a force is applied to it.

Burette: a piece of glassware that measures an accurate volume in millilitres.

Carboxylic acid: group of organic compounds with a COOH functional group.

Catalyst: a substance that provides an alternative pathway for a chemical reaction and so lowers that reaction's activation energy, increasing the rate of the reaction without being used up in the process.

Cation: a positive ion.

Charge separation: where electrons are unequally shared between two atoms or ions creating a positive and negative end in a bond.

Chemical equilibrium: a reversible reaction where the rate of the forward reaction is equal to the rate of the backwards reaction.

Collision theory: the idea that particles must hit other particles with enough force and the correct orientation in order for a chemical reaction to occur.

Compound: a substance made from two or more elements that are chemically bonded to each other.

Concentrated acid: an acidic solution with many acid particles present in it.

Concentration: the amount of particles per unit of volume in $molL^{-1}$ or gL^{-1}.

Conduct: the ability for a substance to allow a current or heat to pass through it.

Conjugate acid: part of a pair of substances that will donate a proton.

Conjugate base: part of a pair of substances that will accept a proton.

Corrosive: a substance that can cause chemical damage to another substance.

Covalent bond: a connection between two atoms where electrons are shared.

Covalent network solid: a substance made from atoms that are connected covalently.

Current: where charged particles travel through a substance.

Delocalised: free to move around.

Delta: slightly.

Dilute acid: an acidic solution with few acid particles present in solution.

Dissociate: when a solute separates into ions as it dissolves.

Dissolve: when the forces of attraction between the solvent and solute are stronger than the forces within the solute.

Double bond: two bonding pairs of electrons between two atoms.

Ductile: the ability for a substance to be drawn into wires.

Dynamic equilibrium: an equilibrium which is constantly in motion.

ISBN: 9780170260107

Electricity: the flow of charged particles through a conductor.

Electron: a negative sub-atomic particle found outside the nucleus of an atom.

Electron configuration: shows how the electrons are arranged in an atom.

Electrostatic attraction: a bond between two oppositely charged particles.

Element: a substance made from one type of atom.

Electron dot diagram: a diagram showing how the electrons are arranged in a molecule.

Electronegativity: the ability for an atom to attract a bonding pair of electrons towards itself in a covalent bond.

Empirical formula: the lowest ratio of atoms found in a molecule.

Endothermic reaction: a reaction where energy is lost from the surroundings in order for bonds to be broken.

Enthalpy: the heat content of a substance.

Equilibrium constant: K is the value of how much product to reactant an equilibrium system has at the equilibrium point at a given temperature.

Exothermic reaction: a reaction where energy is gained by the surroundings as energy is released from bonds being formed.

Flammable: a substance that is easily burnt.

Functional group: an atom or group of atoms responsible for the reactivity of an organic molecule.

Geometric isomer: two alkenes which have the same molecular formula but due to the lack of rotation about the double bond form two different structural isomers.

Group: a column of the periodic table; similar nature of the elements found within it.

Haloalkane: a group of organic compounds with a halogen attached (F, Cl, Br or I).

Hydronium ion: H_3O^+ an ion formed between the reaction of water and a proton.

Ion: an atom or group of atoms that have lost or gained electrons.

Ionic bond: a bond between two oppositely charged ions.

Intermolecular bond: a bond between molecules due to the attraction between them.

Intramolecular bond: a bond within a molecule between the atoms, otherwise known as a covalent bond.

Irritant: a substance that causes discomfort to a part of the human body.

Kinetic energy: the energy of a moving object or particle.

K_w: the equilibrium of water constant which always has the value of 1×10^{14} at 25 °C.

Le Chatelier's principle: any change made to an equilibrium system will make the equilibrium shift in order to correct that change.

Lewis diagram: a diagram showing how the valence electrons are arranged in a molecule.

Linear: a molecular shape with a bond angle of 180°.

Lone pair: in a molecule a pair of electrons that are not part of a bond between two atoms.

Lubricant: a substance that is used to minimise friction.

Malleable: a substance that can be beaten into sheets.

Markovnikov's rule: the carbon that is richer in hydrogen will get richer in an addition reaction.

Mass number: the number of protons and neutrons found in an atom.

Matter: any physical substance that takes up space and has a mass.

Metallic bond: a bond between metal nuclei and their valence electrons.

Melting point: the temperature at which a substance changes from a solid state to a liquid state.

Mixture: a substance made from two or more chemical elements or compounds that are not chemically bonded.

Molecular formula: shows the actual numbers of atoms found in a molecule.

Molecule: a neutral group of atoms that are chemically bonded.

Molecular solid: a solid made from molecules.

Molten: a liquid substance.

Neutral: a substance with no charge, it is neither positive nor negative.

Neutralisation: a reaction between an acid and a base forming a salt and water, both of which will be neutral.

Neutron: a neutral sub-atomic particle in the nucleus of an atom.

Non-polar: a substance that has no charge separation within its bonds.

Non-polar covalent bond: a covalent bond between two

ISBN: 9780170260107

atoms where there is an even sharing of the electrons in the bond.

Non-polar solvent: a substance that will dissolve non-polar substances.

Nucleus: the central part of an atom where the protons and neutrons are.

Organic chemistry: the study of hydrocarbons and their derivatives.

Organic compound: a compound made from a carbon backbone with hydrogen atoms attached.

Oxidation reaction: a reaction where oxygen is gained or electrons are lost from a substance.

Particle: a type of substance: ion, atom, molecule.

Period: a row of the periodic table.

pH: the negative log of the concentration of hydrogen ions present in a solution.

Plane: a dimension in space.

Polar: a substance that has a charge separation within its bonds.

Polar covalent bond: a covalent bond between two atoms where there is an uneven sharing of the electrons in the bond.

Polar solvent: a substance that allows charged particles (either polar substances or ionic substances) to dissolve in it.

Polymerisation reaction: a reaction where many alkenes are added on to each other to make a long chain.

Pressure: the force per unit area that is exerted on or against an object or chemical.

Proton: a positive sub-atomic particle found in the nucleus of an atom.

Reduction reaction: a reaction where electrons are gained by a substance.

Region of negative charge: either a lone pair or a bonding pair of electrons.

Repel: the ability for two like-charged ions or sub-atomic particles to move apart from one another.

Repulsion: when two like-charged ions or sub-atomic particles move apart from one another.

Reversible reaction: a reaction which can go back from the products to the reactants.

Saturated molecule: an organic molecule with the maximum number of bonds attached, and having no double or triple bonds between carbon atoms.

Solubility: the ability of a substance to dissolve in a solvent.

Strong acid: an acid that completely dissociates in water.

Strong base: a base that completely dissociates in water.

Structural formula: the arrangement in space of atoms in an organic molecule.

Structural isomers: compounds with the same molecular formula but a different arrangement in space.

Substitution reaction: a reaction where an atom or group of atoms is exchanged for another atom or group.

Successful collision: a collision that has enough energy behind it in order for two reactants to become products.

Surface area: the total area of the outside layer or surface of an object or chemical.

Symmetrical: a molecular shape that has planes of symmetry in it.

Temperature: how much heat energy an object has in °C or K (Kelvin).

Tetrahedral: shape with four regions of negative charge about a central atom; all four are bonding pairs of electrons with a bond angle of 109°.

Trigonal planar: shape with three regions of negative charge about a central atom; all three are bonding pairs of electrons with a bond angle of 120°.

Trigonal pyramid: shape with four regions of negative charge about a central atom; 3 bonding pairs and one lone pair of electrons with a bond angle of 109°.

Unsaturated molecule: an organic molecule with a double or a triple bond, or multiple double or triple bonds.

Valence shell: the outermost layer of electrons in an atom.

Valency: the number of electrons lost or gained by an ion.

VSEPR theory: Valence Shell Electron Repulsion theory is the idea that electrons repel each other to be as far apart as possible in order for a molecular shape to be formed.

Weak acid: an acid that only partly dissociates in water.

Weak base: a base that only partly dissociates in water.

ISBN: 9780170260107

PERIODIC TABLE

Key	
1	Atomic numer
H	Symbol
Hydrogen	Name
1.008	Atomic Weight

Halogen
Nonmetal
Alkali metal
Alkaline earth metal
Actinide
Transition metal
Other metal
Metalloid
Lanthanide
Noble gases

	1	2	3	4	5	6	7	8	9	10	11	12	13	14	15	16	17	18
1	1 H Hydrogen 1.008																	2 He Helium 4.0026
2	3 Li Lithium 6.94	4 Be Beryllium 9.01218											5 B Boron 10.81	6 C Carbon 12.011	7 N Nitrogen 14.007	8 O Oxygen 15.999	9 F Fluorine 18.9984	10 Ne Neon 20.1797
3	11 Na Sodium 22.98976	12 Mg Magnesium 24.305											13 Al Aluminium 26.9815	14 Si Silicon 28.085	15 P Phosphorus 30.97376	16 S Sulfur 32.06	17 Cl Chlorine 35.45	18 Ar Argon 39.948
4	19 K Potassium 39.0983	20 Ca Calcium 40.078	21 Sc Scandium 44.95591	22 Ti Titanium 47.867	23 V Vanadium 50.9415	24 Cr Chromium 51.9961	25 Mn Manganese 54.93804	26 Fe Iron 55.845	27 Co Cobalt 58.93319	28 Ni Nickel 58.6934	29 Cu Copper 63.546	30 Zn Zinc 65.38	31 Ga Gallium 69.723	32 Ge Germanium 72.63	33 As Arsenic 74.9216	34 Se Selenium 78.96	35 Br Bromine 79.904	36 Kr Krypton 83.798
5	37 Rb Rubidium 85.4678	38 Sr Strontium 87.62	39 Y Yttrium 88.9058	40 Zr Zirconium 91.224	41 Nb Niobium 92.9063	42 Mo Molybdenum 95.96	43 Tc Technetium (98)	44 Ru Ruthenium 101.07	45 Rh Rhodium 102.9055	46 Pd Palladium 106.42	47 Ag Silver 107.8682	48 Cd Cadmium 112.411	49 In Indium 114.818	50 Sn Tin 118.71	51 Sb Antimony 121.76	52 Te Tellurium 127.6	53 I Iodine 126.9044	54 Xe Xenon 131.293
6	55 Cs Caesium 132.9054	56 Ba Barium 137.327	56-71 Lanthanide	72 Hf Hafnium 178.49	73 Ta Tantalum 180.9478	74 W Tungsten 183.84	75 Re Rhenium 186.207	76 Os Osmium 190.23	77 Ir Iridium 192.217	78 Pt Platinum 195.084	79 Au Gold 196.9665	80 Hg Mercury 200.59	81 Tl Thallium 204.38	82 Pb Lead 207.2	83 Bi Bismuth 208.9804	84 Po Polonium (209)	85 At Astatine (210)	86 Rn Radon (222)
7	87 Fr Francium (223)	88 Ra Radium (226)	89-103 Actinide	104 Rf Rutherfordium (267)	105 Db Dubnium (268)	106 Sg Seaborgium (269)	107 Bh Bohrium (270)	108 Hs Hassium (269)	109 Mt Meitnerium (278)	110 Ds Darmstadtium (281)	111 Rg Roentgenium (281)	112 Cn Copernicium (285)	113 Uut Ununtrium (286)	114 Fl Flerovium (289)	115 Uuq Ununpentium (288)	116 Lv Livermorium (293)	117 Uus Ununseptium (294)	118 Uuo Ununoctium (294)

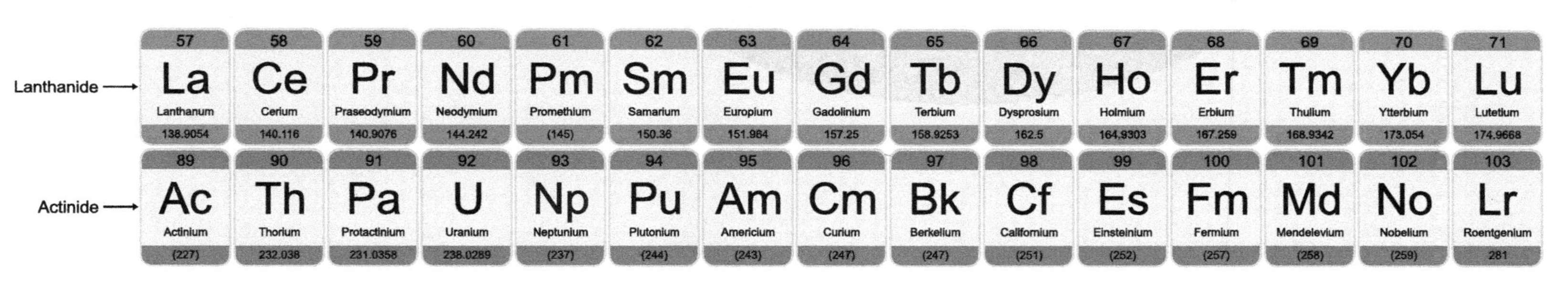

Lanthanide →	57 La Lanthanum 138.9054	58 Ce Cerium 140.116	59 Pr Praseodymium 140.9076	60 Nd Neodymium 144.242	61 Pm Promethium (145)	62 Sm Samarium 150.36	63 Eu Europium 151.964	64 Gd Gadolinium 157.25	65 Tb Terbium 158.9253	66 Dy Dysprosium 162.5	67 Ho Holmium 164.9303	68 Er Erbium 167.259	69 Tm Thulium 168.9342	70 Yb Ytterbium 173.054	71 Lu Lutetium 174.9668
Actinide →	89 Ac Actinium (227)	90 Th Thorium 232.038	91 Pa Protactinium 231.0358	92 U Uranium 238.0289	93 Np Neptunium (237)	94 Pu Plutonium (244)	95 Am Americium (243)	96 Cm Curium (247)	97 Bk Berkelium (247)	98 Cf Californium (251)	99 Es Einsteinium (252)	100 Fm Fermium (257)	101 Md Mendelevium (258)	102 No Nobelium (259)	103 Lr Roentgenium 281

ANSWERS

ISBN: 9780170260107

Chapter One

Pre-test

Section 1: Periodic table

1 They have the same number of valence electrons.
2 Group 1 is called the alkali metals and group 18 is called the noble or inert gases.
3 Period
4 Group
5 Group 2 - +2 and group 17 - -1

Section 2: Atomic structure

1

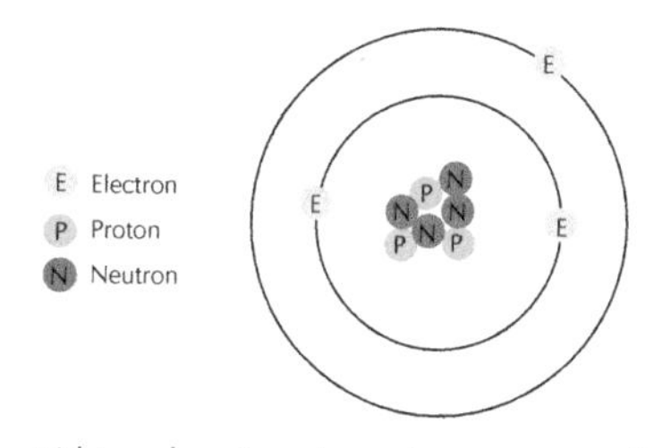

2 Lithium has 3 protons, 4 neutrons and 3 electrons.
3 2, 1

Section 3: Ions

1 a Calcium carbonate
b Aluminium hydroxide
c Potassium nitrate
2 a $Ca(NO_3)_2$
b $Al_2(SO_4)_3$
c NaOH

1.1 – Atomic structure

2 Fluorine

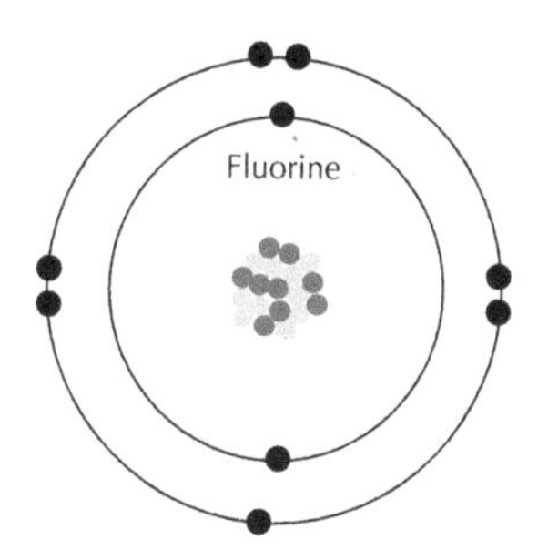

3 Aluminium

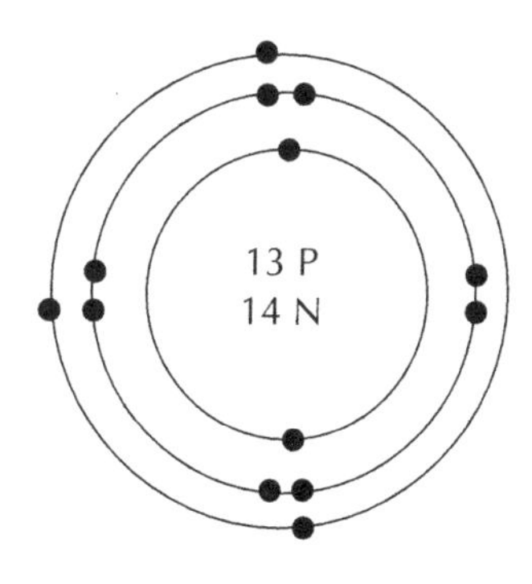

4 Neon

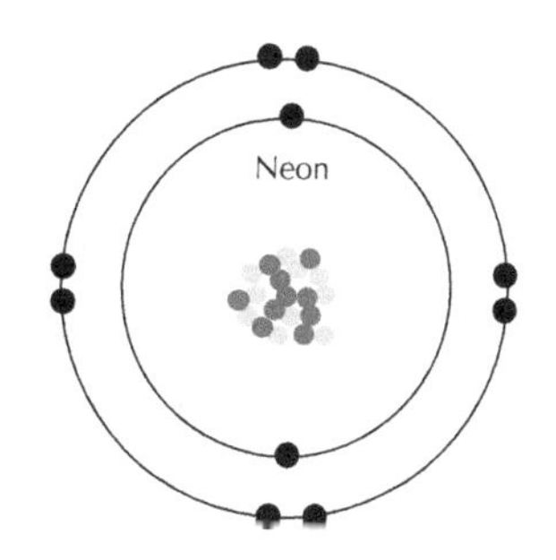

1.2 – Electron arrangements

Element/Ion symbol	Atomic number	Electron configuration
H	*1*	*1*
He	2	2
Li	3	2, 1
Be	4	2, 2
B	5	2, 3
C	6	2, 4
N	7	2, 5
O	8	2, 6
F	9	2, 7
Ne	10	2, 8
Na	11	2, 8, 1
Mg	12	2, 8, 2
Al	13	2, 8, 3
Si	14	2, 8, 4
K	19	2, 8, 8, 1
Ca	20	2, 8, 8, 2
F^-	9	2, 8
Li^+	3	2
Mg^{2+}	12	2, 8
S^{2-}	16	2, 8, 8

1.3 – Ions

	Valency					
	1		2		3	
Metals	Lithium	Li^+	Magnesium	Mg^{2+}	Aluminium	Al^{3+}
	Sodium	Na^+	Calcium	Ca^{2+}	Iron (III)	Fe^{3+}
	Potassium	K^+	Copper (II)	Cu^{2+}		
	Silver	Ag^+	Zinc	Zn^{2+}		
	Copper (I)	Cu^+	Iron (II)	Fe^{2+}		
			Lead	Pb^{2+}		
			Barium	Ba^{2+}		
Non-metals	Fluoride	F^-	Oxide	O^{2-}		
	Chloride	Cl^-	Sulfide	S^{2-}		
	Bromide	Br^-				
	Hydrogen	H^+				
Groups of atoms	Hydroxide	OH^-	Carbonate	CO_3^{2-}	Phosphate	PO_4^{3-}
	Nitrate	NO_3^-	Sulfate	SO_4^{2-}		
	Ammonium	NH_4^+				
	Bicarbonate/ hydrogen carbonate	HCO_3^-				

ISBN: 9780170260107

1.4 – Ionic formulas

Ions	Chloride Cl^-	Hydroxide OH^-	Nitrate NO_3^-	Sulfate SO_4^{2-}	Sulfide S^{2-}	Carbonate CO_3^{2-}	Phosphate PO_4^{3-}
Sodium Na^+	$NaCl$	$NaOH$	$NaNO_3$	Na_2SO_4	Na_2S	Na_2CO_3	Na_3PO_4
Ammonium NH_4^+	NH_4Cl	NH_4OH	NH_4NO_3	$(NH_4)_2SO_4$	$(NH_4)_2S$	$(NH_4)_2CO_3$	$(NH_4)_3PO_4$
Potassium K^+	KCl	KOH	KNO_3	K_2SO_4	K_2S	K_2CO_3	K_3PO_4
Calcium Ca^{2+}	$CaCl_2$	$Ca(OH)_2$	$Ca(NO_3)_2$	$CaSO_4$	CaS	$CaCO_3$	$Ca_3(PO_4)_2$
Magnesium Mg^{2+}	$MgCl_2$	$Mg(OH)_2$	$Mg(NO_3)_2$	$MgSO_4$	MgS	$MgCO_3$	$Mg_3(PO_4)_2$
Aluminium Al^{3+}	$AlCl_3$	$Al(OH)_3$	$Al(NO_3)_3$	$Al_2(SO_4)_3$	Al_2S_3	$Al_2(CO_3)_3$	$AlPO_4$
Iron (II) Fe^{2+}	$FeCl_2$	$Fe(OH)_2$	$Fe(NO_3)_2$	$FeSO_4$	FeS	$FeCO_3$	$Fe_3(PO_4)_2$
Iron (III) Fe^{3+}	$FeCl_3$	$Fe(OH)_3$	$Fe(NO_3)_3$	$Fe_2(SO_4)_3$	Fe_2S_3	$Fe_2(CO_3)_3$	$FePO_4$
Lead Pb^{2+}	$PbCl_2$	$Pb(OH)_2$	$Pb(NO_3)_2$	$PbSO_4$	PbS	$PbCO_3$	$Pb_3(PO_4)_2$
Copper (I) Cu^+	$CuCl$	$CuOH$	$CuNO_3$	Cu_2SO_4	Cu_2S	Cu_2CO_3	Cu_3PO_4
Copper (II) Cu^{2+}	$CuCl_2$	$Cu(OH)_2$	$Cu(NO_3)_2$	$CuSO_4$	CuS	$CuCO_3$	$Cu_3(PO_4)_2$

1.5 – Ionic and metallic bonding

	Ionic bonding	**Metallic bonding**
Type of structure? Giant or Molecular	Giant	Giant
Type of attraction	Electrostatic	Electrostatic
Type of particles present in structure – ions, atoms or molecules	Ions	Atoms
Type of bond	Ionic	Metallic
Diagram of bond		
Examples of where this type of bonding occurs	$NaCl$, $Mg(OH)_2$, $Al_2(CO_3)_3$	Mg, Al, Zn

Checkpoint 1

1 A compound is two or more elements that are chemically bonded, whereas a mixture is two or more chemicals that are placed together but are not chemically bonded.

2 An ion is an atom that has lost or gained electrons, whereas a molecule is two or more atoms that are chemically bonded.

3 **a** B *2,3*
b Al *2,8, 3*
c F^- *2, 8*
d Mg^{2+} *2, 8*

4 **a** CuF ***Copper (I) fluoride***
b $Fe(OH)_3$ ***Iron (III) hydroxide***
c $MgCO_3$ ***Magnesium carbonate***
d Al_2O_3 ***Aluminium oxide***

5 **a** Sodium chloride ***NaCl***
b Lead carbonate ***$PbCO_3$***
c Zinc phosphate ***$Zn_3(PO_4)_2$***
d Iron (II) hydrogencarbonate ***$Fe(HCO_3)_2$***

6 An ionic bond is made between two ions of different charges. It forms as after a chemical reaction one atom gives an electron(s) to another atom to gain a full outer shell. These ions now share an electrostatic attraction as they are oppositely charged.

ISBN: 9780170260107

7 A metallic bond is the electrostatic attraction between the nuclei of the metal atoms and the delocalised valence electrons.

1.6 – Lewis diagrams

1 F_2	:F:F: (three lone pairs on each F)	**3** CO	:C ═ O:	**5** CH_4	H—C—H with H above and below C	**7** CH_3Cl	H—C—Cl: with H above and below C (three lone pairs on Cl)
2 HCl	H—Cl: (three lone pairs on Cl)	**4** NH_3	H—N—H with H below N (one lone pair on N)	**6** CF_4	:F:—C—:F: with :F: above and below C (three lone pairs on each F)	**8** BF_3 Don't forget B is an exception!	:F—B—F: with :F: below B (three lone pairs on each F)

1.7 – Shapes of molecules

Molecule	Lewis diagram	Shape and angle	Shape diagram
N_2	:N ═ N:	Linear, 180°	N ═ N
CO_2	O = C = O (two lone pairs on each O)	Linear, 180°	O = C = O
H_2S	H : S : H (two lone pairs on S)	Bent, 109°	S bonded to two H at a bent angle (two lone pairs on S)
SO_3	:O: double-bonded above S; :O: and :O: single-bonded below S	Trigonal planar, 120°	:O: double-bonded above S; :O: and :O: single-bonded below S
CCl_4	:Cl:—C—:Cl: with :Cl: above and below C (three lone pairs on each Cl)	Tetrahedral, 109°	Tetrahedral C with four Cl
CH_3Cl	H—C—Cl: with H above and below C (three lone pairs on Cl)	Tetrahedral, 109°	Tetrahedral C with three H and one Cl

1.8 – Explaining shapes of molecules

1 H—N—H with H below N (one lone pair on N)   :F—B—F: with :F: below B (three lone pairs on each F)

In ammonia there is one lone pair and three bonded pairs surrounding the central nitrogen atom, so there are four areas of negative charge in total. It will have a bond angle of 109° due to the repulsion between the areas of negative charge which repel each other as far apart as possible giving a trigonal pyramid shape.

In BF_3 there are no lone pairs and three bonding pairs of electrons surrounding the central boron atom, so three areas of negative charge in total. It will have a bond angle of 120° due to the repulsion between the areas of negative charge which repel each other as far apart as possible giving a trigonal planar shape.

The molecules have different bond angles and shapes as they have different amounts of regions of negative charge surrounding the central atoms.

2

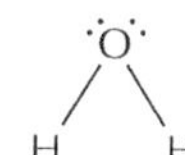

O = C = O (two lone pairs on each O)

In water there are two lone pairs and two bonded pairs of electrons surrounding the central oxygen atom, so four areas of negative charge in total. It will have a bond angle of less than 109° due to the repulsion of these four areas of negative charge which repel each other as far apart as possible giving a bent shape.

In carbon dioxide there are no lone pairs and four bonded pairs of electrons arranged into two double bonds around the central carbon atom, so two areas of negative charge in total. It will have a bond angle of 180° due to the repulsion of these two areas of negative charge which repel each other as far apart as possible giving a linear shape.

Since water has two more areas of negative charge than carbon dioxide it has a smaller bond angle and a different shape to carbon dioxide.

1.9 – Polar, non-polar or ionic

Bond	Polar	Non-polar	Ionic
F_2	X	Yes	X
NaCl	X	X	Yes
HBr	Yes	X	X
O_2	X	Yes	X
MgS	X	X	Yes
CCl_4	Yes	X	X
HCl	Yes	X	X
NH_3	Yes	X	X
Al_2O_3	X	X	Yes
Fe_2S_3	X	X	Yes

ISBN: 9780170260107

1.10 – Polar or non-polar molecules

Molecule	Does it contain polar bonds?	Is the shape symmetrical?	Is the molecule polar or non-polar?
O_2	No	Yes	Non-polar
NH_3	Yes	No	Polar
CF_4	Yes	Yes	Non-polar
CH_3Cl	Yes	No *(Even though the shape is tetrahedral it contains different bonds (C – H and C – Cl) making it not symmetrical in terms of charge.)*	Polar
H_2O	Yes	No	Polar
PCl_3	Yes	No	Polar
SO_2	Yes	No	Polar
SO_3	Yes	Yes	Non-polar

Investigation 1 – Polar or non-polar substances

Substance	Observation	Polar or non-polar?
Water	The water stream moves towards the balloon.	Polar
Cyclohexane	No change to stream flow.	Non-polar
Ethanol	The ethanol stream moves towards the balloon.	Polar
Kerosene	No change to stream flow.	Non-polar

1 As they are also charged and so they move to line up opposite charges with the balloon. So in water's case the hydrogen atoms rotate to face the negatively charged balloon and so the water bends towards it as it is now attracted to it. The hydrogen attached to the oxygen in ethanol also does this for the same reason.

2 The other substances are non-polar and so do not have a charge separation and so are not attracted to the charged balloon.

Checkpoint 2 – Covalent bonds, shape and polarity

1 In an ionic bond there is a permanent charge separation between the two ions involved in the bond, where one has permanently given up an electron to the other ion. In a polar covalent bond there is a difference in electronegativity between the atoms in the bond, however it is not large enough for a permanent separation of charge so instead the electrons are shared but shared unequally and so spend more time surrounding one of the atoms in the bond than the other.

2 NaF contains ionic bonds and it is made up of ions. The valence electron in sodium is permanently given to F, making it Na^+ and Fluorine F^- (fluoride). In F_2 there are non-polar covalent bonds between the fluorine atoms and the atoms share the bonding pair of electrons equally as there is no difference in electronegativity between the two atoms.

3

Molecule	Lewis diagram	Shape name and diagram	Polar or non-polar molecule
CH_3F	H–C(H)(H)–F:	Tetrahedral	Polar
BH_3	H–B(H)–H	Trigonal planar	Non-polar
Cl_2O	:Cl–O–Cl:	**Bent**	Polar
Br_2	:Br : Br:	Linear Br - Br	Non-polar

4

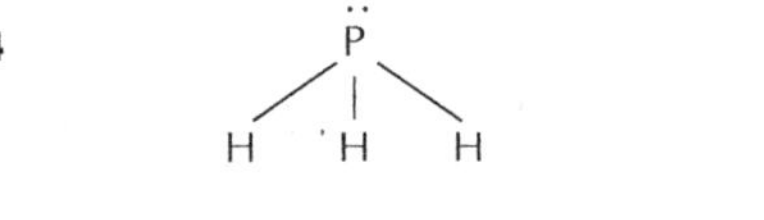

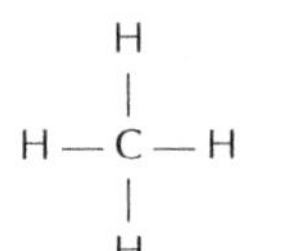

PH_3 has four regions of negative charge surrounding its central Phosphorus atom, one lone pair and three bonding pairs of electrons, giving it a shape of trigonal pyramid. These four regions of negative charge repel each other so they are as far apart as possible giving a bond angle of 109°.

CH_4 has also got four regions of negative charge surrounding its central carbon atom, however all four are bonding pairs, giving it a tetrahedral shape. These four regions also repel each other as far apart as possible giving a bond angle of 109°.

Since there are only three bonding pairs of electrons in PH_3 rather than the four in CH_4 they have different shapes, despite both molecules having four regions of negative charge.

Investigation 2 – Forces between particles

1 Iodine, sulfur and ice all melt. The fact that these can melt tells us they have weak forces of interaction between their molecules which require less energy to break than in the other substances. This force of attraction is the intermolecular forces which hold the molecules together, not the covalent bonds in the molecules themselves.

2 When a substance melts, the forces of attraction between the molecules, ions or atoms (depending on the substance).are broken. This is because the kinetic energy of the particles is increased until they eventually can no longer be held together by their bonds.

1.11 – The properties of solids

1 Aluminium is a metallic solid with a strong electrostatic attraction between the metal nuclei and the valence electrons. Aluminium sulfate is an ionic solid with a strong electrostatic attraction between the positively charged Al^{3+} ions and the negatively charged SO_4^{2-} ions. They both have such high melting points because it takes a lot of energy to overcome the metallic or ionic bonds between the particles in these substances.

ISBN: 9780170260107

2 Aluminium conducts electricity when solid or molten as it is made up of an electrostatic attraction between the positive metal nuclei and the delocalised valence electrons. As the valence electrons are delocalised they are free to move without breaking the strong metallic bonds to conduct a current. Aluminium sulfate can't conduct when solid as it is made from positive aluminium ions and negatively charged sulfate ions that are tightly bonded together with ionic bonds. This means the ions are unable to move when in the solid state as they are held together in a fixed lattice structure. When molten or aqueous these ions are then free to move and so conduct current as in order for a current to flow there must be free-moving charged particles.

3 Aluminium is ductile as it is made from a strong electrostatic attraction between the metal nuclei and delocalised valence electrons. This metallic bond, which is strong and non-directional, meaning the electrons and nuclei can move around without breaking the bonds. This property means it can be drawn into wires or means that metals are ductile. Aluminium sulfate is brittle, as when it is struck with a blunt object the ionic lattice becomes disrupted. Ionic solids like aluminium sulfate contain ionic bonds between the oppositely charged ions present, in this case aluminium and sulfate ions. When the lattice structure is disrupted the ions are no longer lined up positive to negative and can be lined up positive to positive or negative to negative. Like-charged ions repel and so the lattice breaks off where like-charged ions have been forced together. This makes aluminium sulfate brittle as it shatters easily.

1.12 – Properties of solids 2

1 Oxygen is a molecular solid made up of covalent bonds between the oxygen atoms and intermolecular forces between the molecules of oxygen. These weak intermolecular forces are broken easily as they only require a very small amount of energy to break. In contrast zinc has strong metallic bonds between its metal nuclei and valence electrons which require more energy to break apart. It is the same with zinc oxide – it has strong ionic bonds between the ions which are strongly attracted to each other in electrostatic attraction and require more energy to break apart than oxygen.

2 Sodium chloride is an ionic solid made from sodium and chloride ions that are attracted to each other because they are oppositely charged. As the ions are charged they are attracted to the charged water molecules and the force of attraction between the water molecules and the ions is strong enough to overcome the ionic bonds.

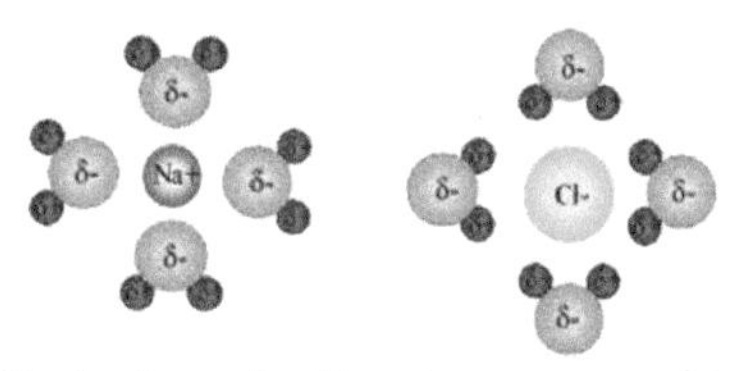

The positively charged sodium ions are attracted to the slightly negatively charged oxygen ends of the water molecules. The negatively charged chloride ions are attracted to the slightly positively charged hydrogen ends of the water molecules as shown in the diagram above. In contrast, octane is not charged as it is a non-polar molecule formed by covalent bonds between the carbon and hydrogen atoms (and carbon and carbon atoms) and weak intermolecular bonds between the octane molecules. As it is not a charged molecule, the force of attraction between the water molecules which are polar and the octane molecules is not strong enough to overcome the intermolecular bonds between each of the octane molecules (and water molecules) and so the octane will float on the surface of the water.

Investigation 3 – Solubility and bonding

Substance	Type of bonding	Observation in water	Observation in cyclohexane
potassium iodide	Ionic	Dissolves	No change
Sugar	Intermolecular	Dissolves	No change
Copper sulfate	Ionic	Dissolves	No change
Silica	Covalent	No change	No change
Wax	Intermolecular	No change	Dissolves
Ethanol	Intermolecular	Dissolves	No change
Iodine	Intermolecular	Partially soluble	Dissolves
Graphite	Covalent	No change	No change
Aluminium	Metallic	No change	No change

1 Substances that dissolved in water were **charged and either contained polar covalent bonds or ionic bonds.** Substances that dissolved in cyclohexane were **not charged and contained metallic bonds, non-polar covalent bonds.** Insoluble substances in both water and cyclohexane were **not attracted to either non-polar or polar solvents like covalent network solids or metals.**

2 Because water is a polar covalent solid it is charged. The slightly positive hydrogen atoms form a bond with the negative ion from an ionic substance, and the slightly negative oxygen atoms form a bond with the positive ions from an ionic substance. This force of attraction between the water molecules and the ions is strong enough to overcome the force of attraction between the ions or the ionic bonds.

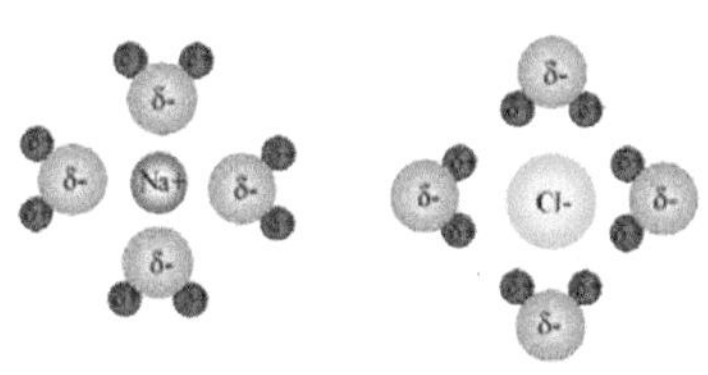

3 These molecules must be polar as well, as then the charged substances will be able to form an attraction strong enough to overcome the forces of attraction between the polar molecules known as the intermolecular forces. Once again the slightly negative end of the polar molecule will be attracted to the slightly positive hydrogen ends in the water, and the slightly positive end in the polar molecule will be attracted to the slightly negative oxygen end of the water molecule.

Investigation 4 – Making models of diamond and graphite

1 The diamond structure appeared to show more strength as each carbon atom in diamond is covalently bonded to four others in a tetrahedral arrangement, whereas in graphite each carbon atom is only connected to three others, leaving weak intermolecular forces between the layers.

2 Diamond is the hardest substance on Earth because each carbon is covalently bonded to four others, in a tetrahedral arrangement which requires a large amount of energy to break apart. In graphite each carbon atom is only covalently bonded to three others, leaving weak intermolecular bonds between the layers. These intermolecular bonds can be broken relatively easily allowing the layers of graphite to slide over one another without breaking the covalent bonds within the layers.

ISBN: 9780170260107

1.13 – Properties of covalent network solids

1. In diamond each carbon atom is covalently bonded to four other carbon atoms, allowing no free valence electrons in carbon to conduct a current as all the valence electrons are incorporated in covalent bonding. In order to conduct a current, a substance must contain delocalised charged particles. In graphite each carbon atom is covalently bonded to three other carbon atoms leaving one valence electron free to conduct a current.
2. Silicon dioxide has a high melting point as it only contains strong covalent bonds between the oxygen and silicon atoms that require a large amount of energy to break, hence the high melting point.

1.14 – NCEA style questions - types of solids

Substance	Type of particle	Attractive forces between particles
Zinc hydroxide, $Zn(OH)_2$	Ions	Ionic
Aluminium oxide, Al_2O_3	Ions	Ionic
Water, H_2O	Molecules	Intermolecular
Oxygen, O_2	Molecules	Intermolecular
Hydrochloric acid, HCl	Molecules	Intermolecular
Octane, C_8H_{18}	Molecules	Intermolecular
Zinc, Zn	Atoms	Metallic
Graphite, C	Atoms	Covalent
Silicon dioxide, SiO_2	Atoms	Covalent
Lead, Pb	Atoms	Metallic

Checkpoint 3 - Summary of the types of solids

Type of solid	Type of particle present in it – ions, atoms or molecules	Bonds present in the structure – ionic, metallic, covalent, intermolecular	Properties of the structure – hard/soft, conductor/insulator, high/low melting point	Diagram of the structure	Examples
Ionic	Ions	Ionic	Hard and brittle Conductor when molten or aqueous High melting point		NaCl, $Zn(OH)_2$, Al_2O_3
Metallic	Atoms	Metallic	Malleable, ductile Conductor High melting point		Zn, Al, Mg, Na
Molecular	Molecules	Covalent, intermolecular	Soft Insulator Low melting point		F_2, Cl_2, HCl, H_2O
Covalent network	Atoms	Covalent, (intermolecular in graphite)	Hard Insulator, except graphite High melting point		Diamond, graphite, silicon dioxide

Investigation 5 – Observing chemical reactions

1. Adding $MgSO_4$ to water, adding NaOH to water and adding Mg to HCl.
2. Adding NH_4Cl to water.
3. The making of bonds.

1.15 – Endothermic or exothermic reactions

1. Exothermic. Reason: The change in enthalpy is a negative value which means heat is released into the surroundings.
2. Exothermic. Reason: The reactants have more energy than the products which means heat has been released into the surroundings.
3. Endothermic. Reason: More energy is going into breaking bonds in carbon dioxide and water than into making the bonds in glucose and oxygen, or energy is absorbed during the reaction.

1.16 – Enthalpy calculations

a Energy = ΔH x mol = 478 x 0.80 = 382.4 = 382 kJ released (3 s.f.)

b Energy = ΔH x mol = 20 x 52 = 1040 kJ released

c mol = mass/Molar mass = 50/(6 x 12 + 12) = 0.595... mol
Energy = ΔH x mol =3920 x 0.595... = 2333.333... = 2330 kJ released (3 s.f.)

d mol = mass/Molar mass =12/(12 + 16 x 2) = 0.2727... mol
Energy = ΔH x mol =469 x 0.2727 = 127.90... = 128 kJ absorbed (3 s.f.)

e mol = mass/Molar mass =3000/(12 x 3 + 8) = 68.18... mol
Energy = ΔH x mol =2220 x 68.18... = 151363... = 151000 kJ released (3 s.f.)

1.17 – Bond energies

1 ΔH = bonds broken – bonds formed
= (4C – H + 2O = O) – (2C = O + 4H – O)
= (4 x 413 + 2 x 494) – (2 x 799 + 4 x 459)
= 2640 – 3434
= - 794 kJmol-1

2 ΔH = bonds broken – bonds formed
= (H – H + Cl - Cl) – (2H - Cl)
= (432 + 242) – (2 x 431)
= 674 – 862
= - 188 $kJmol^{-1}$

ISBN: 9780170260107

3 ΔH = bonds broken – bonds formed
= (4C – H + Cl - Cl) – (3C - H + C – Cl + H - Cl)
= (4 x 413 + 242) – (3 x 413 + 328 + 431)
= 1894 – 1998
= - 104 $kJmol^{-1}$

4 ΔH = bonds broken – bonds formed
= (N ≡ N + 3 x H - H) – (6 x N - H)
= (941 + 3 x 432) – (6 x 391)
= 2237 – 2346
= - 109 $kJmol^{-1}$

5 ΔH = bonds broken – bonds formed
+26.5 = (H - H + I - I) – (2H - I)
+26.5 = (432 + 151) – (2H - I)
- H - I = (26.5 - 583)/2
H - I = 278.25
= + 278$kJmol^{-1}$ (3 s.f.)

Exam-type questions

1 a

Molecule	Cl_2	SF_2	PBr_3
Lewis structure	:Cl : Cl:	:F : S : F:	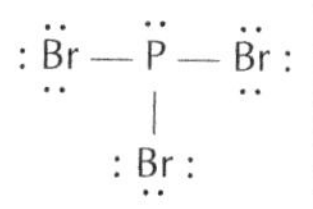 :Br — P — Br: with :Br: below P
Molecule	SO_3	CF_4	F_2O
Lewis structure	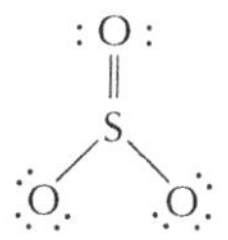 :O: double-bonded to S, S bonded to :O: and :O:	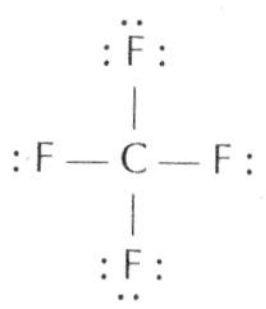 :F: above, :F — C — F:, :F: below	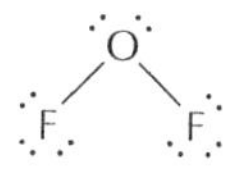 O bonded to F and F

b

Molecule	BeH_2	SO_2
Lewis structure	H : Be : H	 S bonded to O and O

i BeH_2
Shape: Linear
Reason: There are two regions of negative charge about the central Be atom which are both bonded pairs of electrons between the Be and H, these repel each other to be as far apart as possible giving a linear shape with bond angles of 180°.

ii SO_2
Shape: Bent
Reason: There are three regions of negative charge about the central S atom, one double bond to an oxygen, one bonded pair connected to the other O and one lone pair. These regions of negative charge repel each other so they are as far apart as possible giving a bent shape and a bond angle of 120°.

2 a

Molecule	O_2	NCl_3	$CHCl_3$
Lewis structure	O ═ O	 :Cl: above, :Cl : N : Cl:	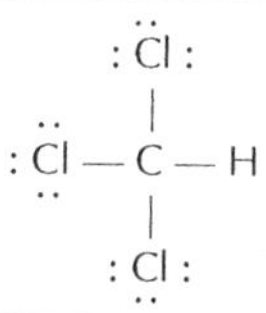 :Cl: above, :Cl — C — H, :Cl: below
Shape	**Linear**	**Trigonal pyramid**	**Tetrahedral**
Shape diagram	O ═ O	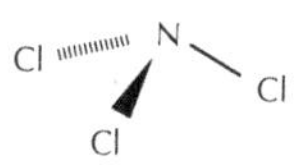 N with Cl, Cl, Cl	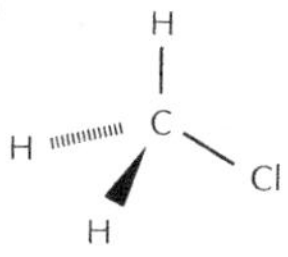 C with H, H, Cl, Cl

b NCl_3 has four regions of negative charge about its central N atom, it has one lone pair of electrons and three bonding pairs of electrons. These repel each other in order to be as far apart as possible giving a bond angle of slightly less than 109° and a trigonal pyramid shape. $CHCl_3$ also has four regions of negative charge about its central C atom, it has four bonding pairs of electrons and no lone pairs of electrons. These regions of negative charge also repel each other to be as far apart as possible giving a bond angle of 109° but a tetrahedral shape.

3 i
a H_2O **polar bonds**
b O_2 **non-polar bonds**

ii Water contains polar bonds as there is a difference in electronegativity between the hydrogen and oxygen atoms as they are not the same atoms. With O being more electronegative than H.
Oxygen does not contain polar bonds but contains non-polar bonds as there is no difference in electronegativity between the two O atoms since they are the same atom.

iii
a CH_4 **non-polar molecule**
b NH_3 **polar molecule**

iv Methane contains polar bonds as there is a difference in electronegativity between the C and the H atoms, with the C being more electronegative than the H. However, as the shape of the methane molecule is symmetrical (it is tetrahedral) the bond dipoles cancel making the molecule overall non-polar.
Ammonia contains polar bonds as there is a difference in electronegativity between the N and H atoms, with the N being more electronegative than the H. As the trigonal pyramid shape of the molecule is asymmetrical due to the lone pair of electrons, these bond dipoles will not cancel out making the molecule polar.

4 a

Molecule	Shape diagram
CCl_3H	Shape: ***tetrahedral*** 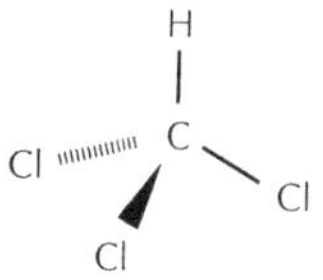 (C with H, Cl, Cl, Cl)
CCl_4	Shape: ***tetrahedral*** 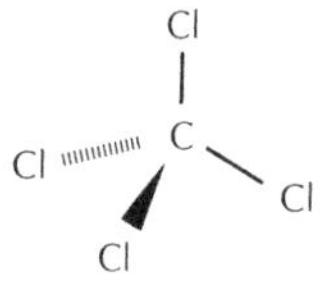 (C with Cl, Cl, Cl, Cl)

b $CHCl_3$ **Polar**
CCl_4 **Non-polar**

$CHCl_3$ contains polar bonds between the carbon and the hydrogen where the carbon is slightly negatively charged as it is the more electronegative atom, and between the carbon and the chlorine where the chlorine is slightly negatively charged as it is the more electronegative atom. These bond dipoles will not cancel out due to the fact that they are unevenly spread about the central atom, making the molecule asymmetrical and making the molecule polar overall.
CCl_4 also contains polar bonds between the carbon and chlorine with the chlorine being slightly negative as it is the more electronegative atom. These bond dipoles do cancel out however, as they are symmetrically arranged about the central C atom, making the molecule non-polar overall.

ISBN: 9780170260107

5 a

Solid	Type of solid	Type of particle present	Attractive forces between particles
Cl_2	*Molecular*	*Molecules*	*Intermolecular*
$SiCl_2$	*Molecular*	*Molecules*	*Intermolecular*
SiO_2	*Covalent network*	*Atoms*	*Covalent*
Graphite	*Covalent network*	*Atoms*	*covalent*

b Graphite is a conductor of electricity because of its structure. It is made up of carbon atoms which are connected to three other carbon atoms, each leaving one free valence electron on each carbon atom which is free to conduct a current (for a substance to conduct a current it must have free moving charged particles). Graphite is used as a lubricant and is soft as it has weak intermolecular forces between the layers of carbon atoms which allow the layers to slide over one another without breaking the strong covalent bonds between the atoms. Silicon dioxide on the other hand is made from silicon atoms bonded to four other oxygen atoms in a tetrahedral arrangement with no free electrons to conduct a current; hence it will not conduct electricity. This structure, however, makes it very strong as the bonds between the atoms are all covalent bonds which require a large amount of energy to break, making it very hard.

6 a

Solid	Type of solid	Type of particle present	Attractive forces between particles
Mg	*Metallic*	*Atoms*	*Metallic*
$MgCl_2$	*Ionic*	*Ions*	*Ionic*

b Magnesium is made up of metal nuclei electrostatically bonded to delocalised valence electrons. These valence electrons are free to move and so can conduct a current, as in order for a substance to conduct a current it must have free moving charged particles. This also allows magnesium to be malleable, as the electrons can move around without breaking the metallic bonds to the metal nuclei, meaning it can be beaten into sheets or bent.

Magnesium chloride, on the other hand, is made up of magnesium and chloride ions which are attracted to each other as they are oppositely charged giving them a strong electrostatic attraction. When solid these ions are arranged in a fixed lattice structure alternating between the positive magnesium ions and the negative chloride ions, which means that despite it having charged particles (the ions) these ions are not free to move to conduct a current. However when molten or aqueous these ions are free to move to conduct a current. Magnesium chloride is not malleable as when it is struck a force is applied to it; the ions rearrange their positions forcing ions of the same charge next to each other. As like charges repel this cleaves the ionic lattice and causes it to be brittle.

7 a Cyclohexane is made up of strong covalent bonds between the carbon and hydrogen atoms and weak intermolecular forces between the molecules. These weak intermolecular bonds require very little energy to break, hence the low melting point. Sodium chloride is made up of sodium and chloride ions which are held together by strong ionic bonds which require a much larger amount of energy to break apart; hence it has a much higher melting point.

b Sodium chloride will dissolve in water as water is polar and so has a stronger force of attraction to the ions than the ions have towards each other. Once the sodium chloride enters water, the slightly positive hydrogen end of the water forms a strong attraction to the negative chloride ions, and the slightly negative oxygen end of the water attracts the positive sodium ions breaking the ionic bonds between the sodium and chloride ions. The attraction between the non-polar cyclohexane and the ions in sodium chloride is not strong enough to overcome the force of attraction between the ions, hence sodium chloride is insoluble in cyclohexane.

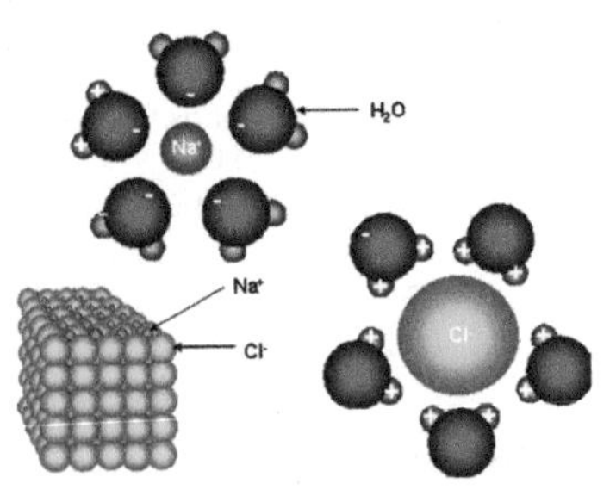

8 a PBr_3 has four regions of negative charge about the central P atom, three bonding pairs and one lone pair. These regions repel each other so they are as far apart as possible forming a trigonal pyramid shape.

BH_3 on the other hand has only got three regions of negative charge about the central atom, three bonding pairs and no lone pairs. These regions repel each other so they are as far apart as possible forming a trigonal planar shape.

As the shape is determined not by number of bonding pairs but by number of regions of negative charge, these two molecules have different shapes.

b PBr_3 has polar bonds as there is a difference in electronegativity between the P and the Br atoms, with the Br being slightly negative as it is the more electronegative atom. These bond dipoles do not cancel each other out as the trigonal pyramid shape is asymmetrical due to the lone pair of electrons creating a polar molecule overall.

BH_3 also has polar bonds between the B and H atoms due to the electronegativity difference between the two atoms, with H being slightly delta negative as it is the more electronegative atom. These bond dipoles do cancel out however, as the trigonal planar shape is symmetrical making the molecule non-polar overall.

9 a Exothermic. Reason: *The change in enthalpy is negative, meaning that heat energy is being lost to the surroundings.*

b ΔH = bonds broken - bonds formed
= (1/2 Cl – Cl + ½ H – H) – (H – Cl)
= (1/2 x 242 + ½ x 432) – 431
= 337 – 431
= - 94.0 $kJmol^{-1}$ (3 s.f.)

c Moles = Mass/Molar Mass = 10/36.5 = 0.2739... mol
Energy = ΔH x moles = 94 x 0.2739 = 25.8 kJ released (3 s.f.)

10 a Exothermic. Reason: *The change in enthalpy is negative, meaning that heat energy is being lost to the surroundings.*

b Moles = Mass/Molar Mass = 6000/30 = 200 mol
Energy = ΔH x moles = 1559.7 x 200 = 312 000 kJ released (3 s.f.)

c ΔH = bonds broken – bonds formed
=(C–C+6xC–H+3.5xO=O)–(6xO–H+4xC=O)
-1559.7 = (347 + 6 x 413 + 3.5 x 494) – (6 x O – H + 4 x 799)
-1559.7 = 1358 – (6 x O – H)
O – H = 486 $kJmol^{-1}$ (3 s.f.)

Chapter Two

Pre-test

Section 1: What is organic?

1. Many plastic products such as ballpoint pens and car battery cases. Petrol, kerosene, fuel, natural gas to name a few!
2. skin cells, rubber, plastic, vitamin D, petrol

ISBN: 9780170260107

3 They are all made from carbon and hydrogen.
4 In chemistry organic means any substance made from carbon and hydrogen atoms. You may have a different definition of this.

Section 2: Types of organic substances

1 Pens, pencil cases, lunchboxes, trays etc.
2 Pens, pencil cases, lunchboxes, trays etc.
3 Methanol, ethanol, propanol... or vodka, wine, beer, rum, gin.

2.1 – Structural and condensed structural formula

	Structural formula	Condensed structural formula
1	H | H — C — O — H | H	CH_3OH
2	H H \ / C=C / \ H H	CH_2CH_2
3	H H H | | | H — C — C — C — H | | | H H H	$CH_3CH_2CH_3$
4	H H H H | | | | H — C=C — C — C — H | | H H	$CH_2CHCH_2CH_3$
5	H | H — C — Cl | H	CH_3Cl
6	H H | | H — C — C — H | | H H	CH_3CH_3
7	H H H | | | H — C=C — C — H | H	CH_2CHCH_3
8	H H \ / C=C / \ H H	CH_2CH_2

2.2 – Labelling the number of carbons

	Prefix
1	Prop
2	Meth
3	Eth
4	Eth
5	Meth
6	But
7	Hept
8	But

2.3 – Labelling the functional group

	Prefix and suffix
1	Propane
2	Aminomethane or methanamine
3	Ethanoic acid
4	Ethene
5	Methane
6	Butanol
7	Chloroheptane
8	Butane

2.4 – Numbering and naming

	Name
1	Butan-1-ol
2	Butanamine or 1-aminobutane
3	Pent-1-ene
4	Pent-2-ene
5	Pent-1-yne
6	Butan-1-ol
7	1-chloroheptane
8	Butan-2-ol

2.5 – Naming branched molecules

	Organic structure	Name
1	H H \ / C=C / \ H—C H H/ \H	Propene
2	CH_3 | $CH_3CHCHCH_3$ | CH_3	2,3-dimethylbutane
3	$CH_3CH(OH)CH(CH_3)CH_2CH_3$	3-methylpentan-2-ol
4	$CH_3CH(Cl)CH(CH_3)CH_2CH_3$	2-chloro-3-methylpentane
5	H_3C CH_3 | | $CH_3CHCCH_2CH_2CH_3$ | CH_3	2,3,3-trimethylhexane
6	OH | H_3C — C — CH_3 | CH_3	2-methylpropan-2-ol
7	CH_3 CH_3 | | H_3C — C — C — CH_3 | | CH_3 CH_3	2,2,3,3-tetramethylbutane

ISBN: 9780170260107

Investigation 1: ***Molecules in three dimensions***

	Structural Formula	Condensed structural formula
1	H | H — C — H | H	CH_4
2	H H | | H — C — C — H | | H H	CH_3CH_3
3	H H \ / C = C / \ H H	CH_2CH_2
4	H — C ≡ C — H	CHCH
5	H H | | H — C — C — O — H | | H H	CH_3CH_2OH
6	H H \ / H — C — C — H / \ H Cl	CH_3CH_2Cl
7	H H | | H H — C — C — N | | H H H	$CH_3CH_2NH_2$
8	O || C / \ H OH	HCOOH
9	H | H — C — H H | H H | | | | H — C — C — C — C — H | | | | H H H H	$CH_3CH(CH_3)CH_2CH_3$

QUESTIONS

1 109°

2 It is always tetrahedral about each individual carbon atom.

2.6 – Molecular and empirical formulas

	Molecular formula	Empirical formula
1	C_3H_8	C_3H_8
2	C_4H_{10}	C_2H_5
3	C_5H_8	C_5H_8
4	$C_4H_{10}O$	$C_4H_{10}O$
5	$C_7H_{15}Cl$	$C_7H_{15}Cl$
6	$C_4H_{10}O$	$C_4H_{10}O$
7	$C_6H_{14}O$	$C_6H_{14}O$
8	$C_6H_{13}Cl$	$C_6H_{13}Cl$

2.7 – Structural isomers

1

```
CH3 — CH2 — CH2 — CH2 — CH2 — CH3
             Hexane

CH3 — CH — CH2 — CH2 — CH3
      |
      CH3
   2 – methylpentane

CH3 — CH2 — CH — CH2 — CH3
             |
             CH3
   3 – methylpentane

         CH3
         |
   CH3 — C — CH2 — CH3
         |
         CH3
   2,2 – Dimethylbutane

   CH3 — CH — CH — CH3
         |    |
         CH3  CH3
   2,3 – Dimethylbutane
```

2 Heptane

```
    H   H   H   H   H   H   H
    |   |   |   |   |   |   |
H — C — C — C — C — C — C — C — H
    |   |   |   |   |   |   |
    H   H   H   H   H   H   H
```

2-methylhexane

```
        H
        |
    H — C — H
    H   |   H   H   H   H
    |   |   |   |   |   |
H — C — C — C — C — C — C — H
    |       |   |   |   |
    H       H   H   H   H
```

3-methylhexane

```
            H
            |
        H — C — H
    H   H   |   H   H   H
    |   |   |   |   |   |
H — C — C — C — C — C — C — H
    |   |   |   |   |   |
    H   H   H   H   H   H
```

2,2-dimethylpentane

```
      CH3
      |
CH3 — C — CH2 — CH2 — CH3
      |
      CH3
```

2,3-dimethylpentane

```
      CH3  CH3
      |    |
CH3 — CH — CH — CH2 — CH3
```


 ISBN: 9780170260107

2,4-dimethylpentane

```
    H   H   H   H   H   H
    |   |   |   |   |   |
H — C — C — C — C — C — C — H
    |   |   |   |   |   |
    H   |   H   |   H   H
     H - C - H H - C - H
         |         |
         H         H
```

2, 2, 3-trimethylbutane

```
          CH3
          |
   CH3 — C — CH — CH3
          |    |
         CH3  CH3
```

2.8 – Geometric isomers

	Condensed structural formula	Does this molecule form geometric isomers?	Structural formulas and names of the isomers if they form
2	$CH_3CH_2CH{=}CH_2$	No	
3	$CH_3C(Cl){=}C(Cl)CH_3$	Yes	Cl, Cl / C=C / CH_3, CH_3 *Cis*-2,3-dichlorobut-2-ene; Cl, CH_3 / C=C / CH_3, Cl *Trans*-2,3-dichlorobut-2-ene
4	HC(Br)=C(Br)H	Yes	Br, Br / C=C / H, H *Cis*-1,2-dibromoethene; H, Br / C=C / Br, H *Trans*-1,2-dibromoethene

Checkpoint 1: Naming and isomerism

a 1 – carboxylic acid; 2 – alkene

```
 1 ( O=C—O—H )
       |
  H    C
   \ //  \
    C     (C)— O — C(=O) — C(H)(H) — H
    |     ||
    C     (C)
   / \\  /   2  \
  H    C          H
       |
       H
```

b 1 – amine; 2 – alkene

```
H3C, H2C, CH2, N, H2C, N, C, H2, O, S, O, H, C, C, C, HC, C, H (circle 2), C, O, H2C, CH3, C, N (HN circle 1), O, C, C, N, CH3, N, C, H2O—CH2—CH3
```

2

	Name	Condensed Structural Formula	Molecular and Empirical Formula	Functional Group
a	2,2-dimethylbutane	$CH_3C(CH_3)_2CH_2CH_3$	C_6H_{14}; C_3H_7	Alkane
b	2-chloropropane	$CH_3CH(Cl)CH_3$	C_3H_7Cl	Haloalkane
c	Ethanoic acid	CH_3COOH	$C_2H_4O_2$; CH_2O	Carboxylic acid
d	2-methylpropan-1-ol	$CH_3CH(CH_3)CH_2OH$	$C_4H_{10}O$	Alcohol

3 Structural isomers are compounds with the same molecular formula but a different arrangement in space, shown in the pictures of but-1-ene and but-2-ene below. They both have the same molecular formula of C_4H_8 but are arranged differently in space.

```
     H   H       H
     |   |       |
 H — C — C — C = C          But-1-ene
     |   |   |   |
     H   H   H   H

     H   H       H
     |   |       |
 H — C — C = C — C — H      But-2-ene
     |       |   |
     H       H   H
```

But-2-ene can form geometric isomers as it has two different groups or atoms attached to the two carbons in the double bond and since there is no free rotation about the double bond, these are two different molecules. Whereas in but-1-ene there are two of the same groups on one carbon atom in the double bond.

```
 CH3     CH3            CH3      H
    \   /                  \   /
     C=C                    C=C
    /   \                  /   \
   H     H                H     CH3

 Cis but-2-ene          Trans but-2-ene
```

2.9 – Substitution reactions of alkanes

2 CH_3CH_2Br

3 $CH_3CH_2CH_2CH_2Cl$ (any of the hydrogens could be substituted so a variety of answers can be given here)

4 $CH_3CH_2CH_2CH_2CH_2Br$

5 $CH_3CH_2CH_2CH_2CH_2CH_2CH_2Cl$

2.10 – Reactions of alkenes

3 $CH_3CH(Cl)CH_2Cl$

4 $CH_3CH(OH)CH_3$, $CH_3CH_2CH_2OH$

5

```
   CH3 H  CH3 H  CH3 H
   |   |  |   |  |   |
(- C — C — C — C — C — C -)n
   |   |  |   |  |   |
   H   H  H   H  H   H
```

6 $CH_3CH_2CH_3$
7 $CH_3CH(Br)CH_2CH_3$
8 $CH_3CH(OH)CH_2CH_3$
9 $CH_3CH(OH)CH(OH)CH_3$
10

```
    H   CH3  H   CH3
    |   |    |   |
(- C — C  — C — C -)n
    |   |    |   |
   CH3  H   CH3  H
```

Experiment 1 – Properties and reactions of alkanes and alkenes
METHOD AND RESULTS

Hydrocarbon	Observations: Add 1 mL of acidified potassium permanganate	Add 1 mL of bromine water	Solubility in water (add a few mLs of water and see if the hydrocarbon dissolves)
To 1 mL of cyclohexane	No reaction	Orange solution to colourless at a slow rate	Insoluble
To 1 mL of cyclohexene	Purple solution to colourless	Orange solution to colourless rapidly occurs	Insoluble

1

```
          OH
     H    |       H
       \  |      /
          C        C — OH
     H2C/   \   /
      |       C
      |       |
     H2C      CH2
        \   /
          C
          H2
```

2 Oxidation
3

```
            H
        H   |   H
    H    \  C   |   Br
      \  /    \ |  /
       C        C
   H /  |       |  \ H
        C        C
   H /    \   /    \ H
        H   C   H
            |
            H
```

4 Substitution
5

```
          Br
          |   H
          |  /
          C
   H2C /     \ C — Br
    |          \ H
    |          |
   H2C         CH2
       \     /
          C
          H2
```

6 Addition
7 They are both non-polar molecules and so the forces of attraction between the water molecules and the intermolecular forces between the cyclohexane molecules are not strong enough to overcome the forces between the cyclohexane and cyclohexane molecules.

Checkpoint 2: Hydrocarbons

1

Name: Ethane
Structural formula: CH_3CH_3

Catalyst: UV light

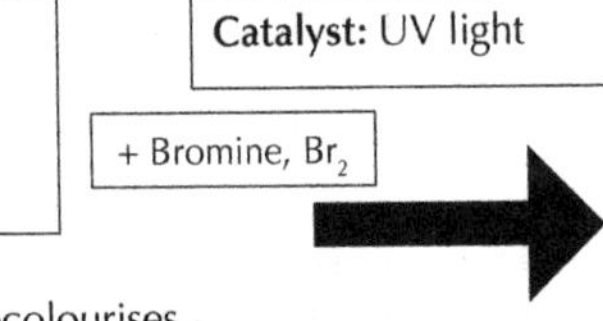

Name of product: bromoethane
Structural formula: CH_3CH_2Br

Orange solution slowly decolourises.

2

Name: Propene
Structural formula: $CH_3CH{=}CH_2$

H^+/MnO_4^- →
Type of reaction: Oxidation
Name of product: Propan-1,2-diol
Structural formula: $CH_3CH(OH)CH_2OH$

Br_2 →
Type of reaction: Addition
Name of product: 1,2-dibromopropane
Structural formula: $CH_3CH(Br)CH_2Br$

H_2/Pt →
Type of reaction: Addition
Name of product: Propane
Structural formula:

```
    H   H   H
    |   |   |
H — C — C — C — H
    |   |   |
    H   H   H
```

Catalyst, heat and pressure →
Type of reaction: Polymerisation
Name of product: Polypropylene
Structural formula:

```
   CH3  H  CH3  H  CH3  H
   |    |  |    |  |    |
(- C  — C — C — C — C — C -)n
   |    |  |    |  |    |
   H    H  H    H  H    H
```

HBr →
Type of reaction: Addition
Name of product: 1-chloropropane + 2-chloropropane
Structural formula: $CH_3CH_2CH_2Cl$ + $CH_3CH(Cl)CH_3$

H^+/H_2O →
Type of reaction: Addition
Name of product: Propan-1-ol and Propan-2-ol
Structural formula: $CH_3CH_2CH_2OH$ + $CH_3CH(OH)CH_3$

 ISBN: 9780170260107

2.11 – Classification and reactions of alcohols

1

Alcohol	$CH_3-CH(OH)-CH_2-CH_3$	$H-C(H)_2-C(H)_2-O-H$	$H-C(H)_2-C(H)_2-C(H)_2-C(H)_2-OH$
Classification	Secondary	Primary	Primary
Alcohol	$CH_3-CH(CH_3)-CH_2-CH_2-OH$	$CH_3-CH_2-C(CH_3)(OH)-CH_2-CH_3$	$H-C(H)_2-C(H)_2-C(H)(OH)-C(H)_2-C(H)_2-H$
Classification	Primary	Tertiary	Secondary

2

a

```
    H   O
    |   ||
H — C — C — O — H
    |
    H
```

b

```
    H   H
    |   |
H — C — C — Cl
    |   |
    H   H
```

c

```
    H   H       H      and     H   H       H
    |   |       |              |   |       |
H — C — C = C — C — H      H — C — C — C = C
    |       |   |              |   |   |   |
    H       H   H              H   H   H   H
```

d No reaction

e

```
    H   H   Cl  H   H
    |   |   |   |   |
H — C — C — C — C — C — H
    |   |   |   |   |
    H   H   H   H   H
```

f

```
    H   H   H   H   H
    |   |   |   |   |
H — C — C = C — C — C — H
    |           |   |
    H           H   H
```

Experiment 2: Solubility and oxidation of alcohols

METHOD AND RESULTS

Alcohol	Observations		
	1 mL of $H^+/Cr_2O_7^{2-}$ added	1 mL of H^+/MnO_4^- added	Solubility in a few mLs of water
To 1 mL of ethanol	Orange solution to green solution	Purple to colourless solutions	Soluble in water

QUESTIONS

1 a CH_3COOH

b CH_3COOH

2 As ethanol is polar due to the large difference in electronegativity between the O and the H in the OH bond, making a polar bond. The bonds between the polar water molecules and the ethanol molecules are strong enough to overcome the intermolecular bonds between the ethanol molecules, therefore it dissolves in water. Water is polar due to the same difference in electronegativity between the O and the H. The O from the ethanol will be slightly negative in charge and will attract the H from the water which is slightly positive and the H from the ethanol (attached to the O) is slightly positive and will form an attraction to the slightly negative O in the water molecules.

2.12 – Classification and reactions of haloalkanes

1

Haloalkane	$CH_3CH_2CH(Br)CH_3$	$H-C(H)_2-C(H)_2-C(H)_2-Cl$	$H-C(H)_2-C(H)(Cl)-C(H)_2-H$
Classification	Secondary	Primary	Secondary
Haloalkane	$H-C(H)_2-C(Br)(CH_3)-C(H)_2-H$	$H_3C-C(Br)(CH_3)-CH_3$	$H-C(H)_2-Cl$
Classification	Tertiary	Tertiary	Primary

2 a

```
    H   H       H      and     H   H       H
    |   |       |              |   |       |
H — C — C = C — C — H      H — C — C — C = C
    |       |   |              |   |   |   |
    H       H   H              H   H   H   H
```

b

```
       NH2
       |
CH3CH2CHCH3
```

c

```
    H
    |
H — C — OH
    |
    H
```

d

```
    H   OH  H
    |   |   |
H — C — C — C — H
    |   |   |
    H   H   H
```

e

```
    H   NH2 H
    |   |   |
H — C — C — C — H
    |   |   |
    H   H   H
```

Checkpoint 3 – Alcohols and haloalkanes

1

Type of reaction: Oxidation
Name: Ethanoic acid
Structural formula: CH_3COOH

Type of reaction: Substitution
Name: Chloroethane
Structural formula: CH_3CH_2Cl

Type of reaction: Substitution
Name: Ethanol
Structural formula: CH_3CH_2OH

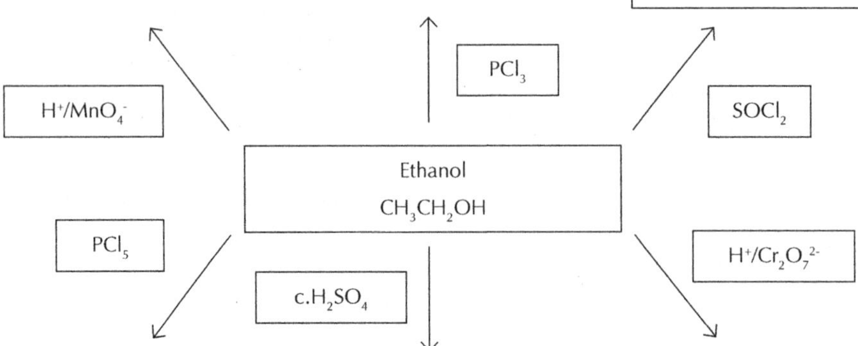

Type of reaction: Substitution
Name: Chloroethane
Structural formula: CH_3CH_2Cl

Type of reaction: Elimination
Name: Ethene
Structural formula: CH_2CH_2

Type of reaction: Oxidation
Name: Ethanoic acid
Structural formula:

2

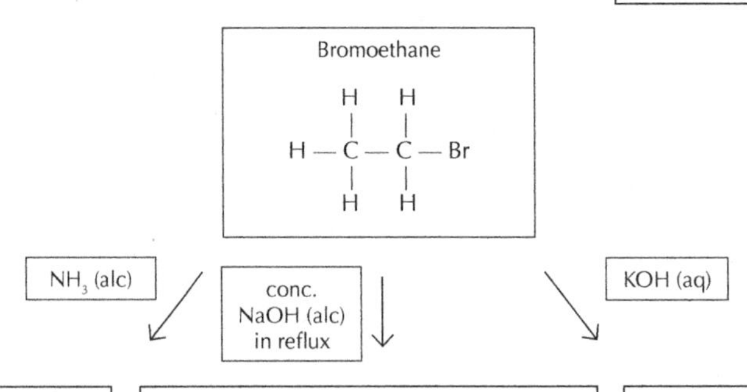

Type of reaction: Substitution
Name: Aminoethane or ethanamine
Structural formula: $CH_3CH_2NH_2$

Type of reaction: Elimination
Name: Ethene
Structural formula: CH_2CH_2

Type of reaction: Substitution
Name: Ethanol
Structural formula: CH_3CH_2OH

3 The propan-1-ol would go from an orange to a green solution, whereas the 1-bromopropane would not change colour and so remain orange. The propan-1-ol would be oxidised to form propanoic acid, shown in the reaction written below.

$$CH_3CH_2CH_2OH \xrightarrow{H^+/Cr_2O_7^{2-}} CH_3CH_2COOH$$

2.13 – Reactions of carboxylic acids

1. $CH_3CH_2COO^- + H_3O^+$
2. $HCOO^- + H_3O^+$
3. $Mg(CH_3CH_2COO)_2 + H_2O$
4. $2LiCH_3CH_2COO + H_2$
5. $2Al(HCOO)_3 + 3H_2O + 3CO_2$
6. $Zn(HCOO)_2 + H_2$

2.14 – Reactions of amines

1. $CH_3CH_2NH_3^+ + OH^-$
2. $CH_3CH_2CH_2CH_2NH_3^+ + OH^-$
3. $CH_3CH_2NH_3^+Cl^- + H_2O$
4. $CH_3CH_2NH_3^+NO_3^- + H_2O$
5. $(CH_3CH_2NH_3^+)_2SO_4^{2-}{}_+ H_2O$

Experiment 3: *Properties of carboxylic acids and amines*

METHOD AND RESULTS

1 Properties of propylamine

	Observations		
	Blue litmus	Red litmus	Carefully smell by using the wafting technique
To 1 mL of propylamine	Stays blue	Goes blue	Strong fishy smell

2 Properties and reactions of ethanoic acid

	Observations				
	Blue litmus	Red litmus	Carefully smell by using the wafting technique	Add a piece of Mg	Add a few calcium carbonate chips
To 1 mL of ethanoic acid	Goes red	Stays red	Strong vinegar smell	Bubbles of gas form	Bubbles of gas form

ISBN: 9780170260107

QUESTIONS

1 $2CH_3COOH + Mg \longrightarrow Mg(CH_3COO)_2 + H_2$

2 $2CH_3COOH + CaCO_3 \longrightarrow Ca(CH_3COO)_2 + CO_2 + H_2O$

3 $CH_3CH_2CH_2NH_2 + H_2O \rightleftharpoons CH_3CH_2CH_2NH_3^+ + OH^-$

4 $CH_3COOH + H_2O \rightleftharpoons CH_3COO^- + H_3O^+$

Checkpoint 4 – Organic chemistry summary

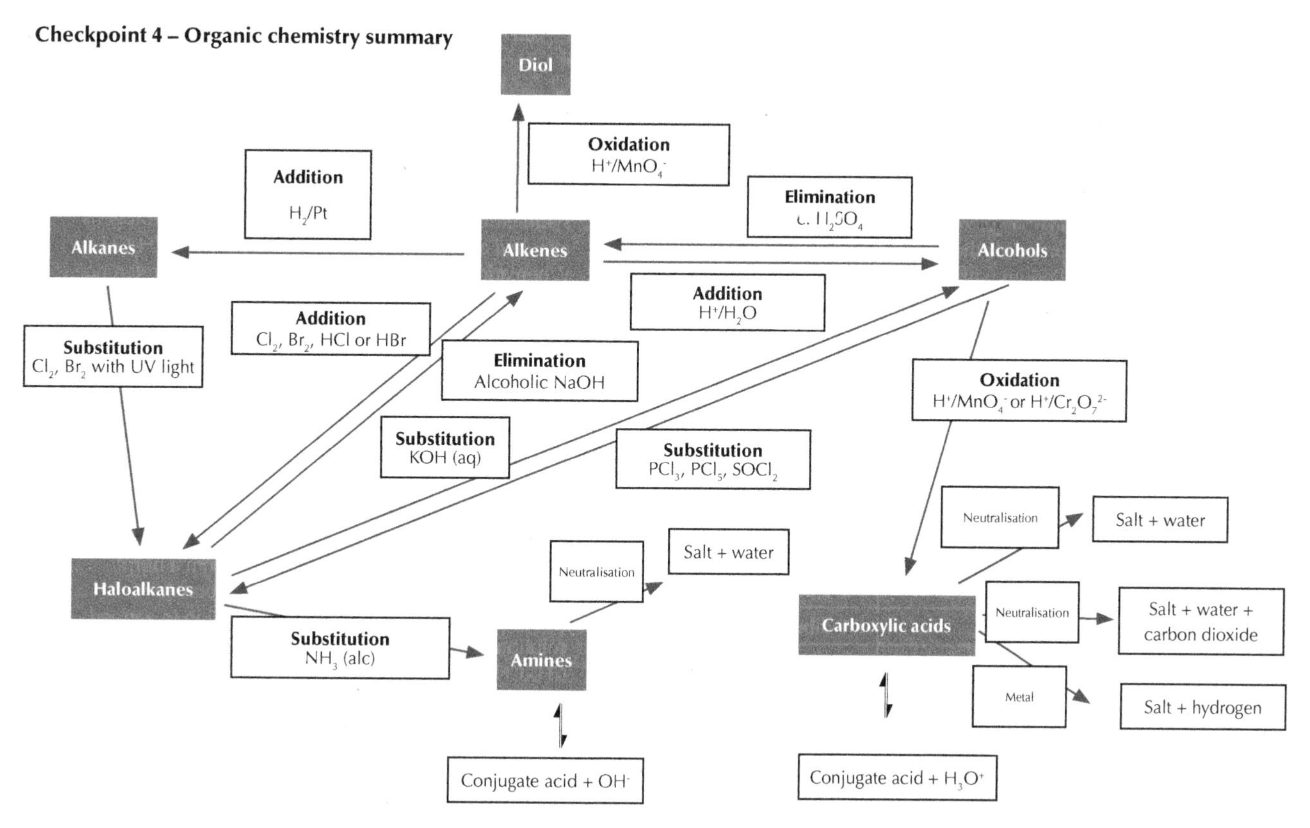

Exam-type questions

1 a

Substance	Structure	Name	Homologous series
A	CH_3 (branch) $CH_3-CH-CH_2-OH$	2-methyl propan-1-ol	Alcohol
B	CH_3 CH_3-C-CH_3 OH	2-methylpropan-2-ol	Alcohol
C	$CH_3CH_2CH_2CH_2NH_2$	1-aminobutane	Amine
D	$CH_3CH(Cl)COOH$	2-chloropropanoic acid	Carboxylic acid
E	$CH_2=C(CH_3)_2$	methylpropene	Alkene

b i Structural isomers are compounds with the same molecular formula but a different arrangement in space. For example C_4H_8 has the following structural isomers:

$CH_3CH_2CH=CH_2$ but-1-ene
$CH_3CH=CHCH_3$ but-2-ene
$CH_2=C(CH_3)_2$ methylpropene

All of the above molecules have the same number of carbons and hydrogens but these atoms are arranged differently in space.

ii Geometric isomers form around a double bond, where each carbon in the double bond has two different groups or atoms attached. This allows two different molecules to form since there is no free rotation about a double bond. Of the above three molecules only but-2-ene can form geometric isomers as it is the only one to have two different groups attached to each carbon in the double bond.

$(CH_3)(H)C=C(CH_3)(H)$ (both CH_3 on the same side) and $(CH_3)(H)C=C(H)(CH_3)$ (CH_3 groups on opposite sides)

2 a

Structural formula	Name
$CH_3-CH_2-CH(CH_3)-C(=O)OH$	2-methylbutanoic acid
$CH_3CH_2-NH_2$	**Ethanamine or aminoethane**
$H_3C-CH(Cl)-CH(Cl)-CH_3$	2,3-dichlorobutane
$H_2C=C(CH_3)-CH_3$ (all H shown)	**2-methylpropene**
$H_3C-CH_2-C(CH_3)(OH)-CH_3$ (all H shown)	2-methylbutan-2-ol

ISBN: 9780170260107

b As aminopropane is a weak base it will turn universal indicator blue, whereas propanoic acid is a weak acid and so will turn universal indicator red.

c Aminopropane is a weak base and so won't react with sodium hydroxide. Whereas propanoic acid is a weak acid and so will completely neutralise sodium hydroxide. This reaction will be slow as propanoic acid only partially dissociates in water to form hydronium ions and its conjugate base, it is the hydronium ions that will react with and neutralise the sodium hydroxide to form salt and water.

$$CH_3CH_2COOH + H_2O \rightleftharpoons CH_3CH_2COO^- + H_3O^+$$

As the reaction proceeds the hydronium ions are used up forcing the equilibrium with water and propanoic acid towards the products and eventually forcing the reaction to go to completion, so the reaction with sodium hydroxide will be slow but will still go to completion. You can tell the reaction has finished when the sodium hydroxide has been neutralised by the propanoic acid by the fact the universal indicator should go from red to green.

$$CH_3CH_2COOH + NaOH \longrightarrow NaCH_3CH_2COO + H_2O$$

3 a

	Structural formula	Name
Molecule A	HC(Br)=CH(Br)	**1,2-dibromoethene**
Molecule B	$Br_2C{=}CH_2$	**1,1-dibromoethene**
Molecule C	$BrCH_2CH_2Br$	**1,2-dibromoethane**

b Structural isomers are molecules with the same molecular formula but a different arrangement in space. Molecules A and B have the same molecular formula $C_2H_2Br_2$ but are arranged differently in space, whereas molecule C has a different molecular formula $C_2H_4Br_2$ and hence is not a structural isomer of A and B.

c Geometric isomers form about a double bond where each carbon in the double bond is connected to two different groups or atoms. This happens because there is no free rotation about a double bond, so two different molecules can form. Molecule A can form geometric isomers as it has a Br and an H attached to each carbon in the double bond, whereas molecule B has 2 Brs attached to one carbon and 2 Hs attached to the other.
The geometric isomers of molecule A:

```
Br      Br          H       Br
  \    /             \     /
   C=C                C=C
  /    \             /     \
H       H          Br       H
```

4 Propene will react with the bromine water, quickly decolourising it from orange to colourless as it reacts to form 1,2-dibromopropane via an addition reaction.
$CH_3CH{=}CH_2 + Br_2 \longrightarrow CH_3CH(Br)CH_2Br$
No other conditions are required.
Propane will only react with bromine water in the presence of UV light, the reaction rate is slow, but the same observation of orange solution to colourless solution will occur.
$CH_3CH_2CH_3 + Br_2 \longrightarrow CH_3CH_2CH_2Br$
Only one hydrogen is exchanged for a bromine atom via a substitution reaction.
Propene will react with acidified permanganate to form propan-1,2-diol via an oxidation reaction. The solution will go from purple to colourless as the permanganate is reduced.
$CH_3CH{=}CH_2 + H^+/MnO_4^- \longrightarrow CH_3CH(OH)CH_2OH$

5 a

```
   H   H   H   H   H   H
   |   |   |   |   |   |
(- C - C - C - C - C - C -)n
   |   |   |   |   |   |
   H   H   H   H   H   H
```

b **i** bromoethane CH_3BrCH_3

ii The double bond in ethene is lost and a Br atom is added on to one carbon in the double bond and a H is added on to the other.

c This is because ethene is a symmetrical alkene, so no matter where you add on the carbons in the double bond the same product is formed. However but-1-ene is not symmetrical as one C in the double has two hydrogens attached and the other has a hydrogen and a CH_2CH_3. When adding to an asymmetrical alkene the preference is for the hydrogen to add to the carbon in the double bond with the most hydrogens already. This is Markovnikov's rule and forms the major product. The minor product therefore will form with the Bromine atom on the end.

$$CH_2{=}CHCH_2CH_3 + HBr \longrightarrow$$
$$CH_3CH(Br)CH_2CH_3 + CH_2(Br)CH_2CH_2CH_3$$
Major   Minor

6 **i** Either use acidified potassium permanganate or use acidified potassium dichromate with heat.

ii With potassium permanganate it goes from a purple solution to a colourless solution, with potassium dichromate it goes from an orange solution to a green solution.

iii $CH_3(CH_2)_4OH + H^+/MnO_4^- \longrightarrow CH_3(CH_2)_3COOH$

7 a Aminoethane would turn red litmus blue and ethanoic acid will turn blue litmus red.

b **i** Neutralisation

ii Because carbon dioxide gas forms.

iii $2CH_3COOH + MgCO_3 \longrightarrow Mg(CH_3COO)_2 + H_2O + CO_2$

iv Because ethanoic acid is a weak acid it only partially dissociates in water to form hydronium ions and its conjugate base.

$$CH_3COOH + H_2O \rightleftharpoons CH_3COO^- + H_3O^+$$

This slows down the reaction rate with magnesium carbonate as the equilibrium must constantly shift as the hydronium ions form, they are not all present at the beginning of the reaction. Eventually, though, the equilibrium is forced to completion hence why the reaction with magnesium carbonate is a complete reaction.

8 With aqueous KOH a substitution reaction occurs where the Cl atom is exchanged with an OH forming propanol.

$$CH_3CH_2CH_2Cl + KOH_{(aq)} \longrightarrow CH_3CH_2CH_2OH$$

With alcoholic ammonia chloropropane forms amino propane as, again, a substitution reaction occurs, however this time the Cl atom is exchanged with an NH_2 group.

$$CH_3CH_2CH_2Cl + NH_{3\,(alc)} \longrightarrow CH_3CH_2CH_2NH_2$$

Finally the reaction with concentrated sodium hydroxide is an elimination reaction where HCl is ejected from the molecule, leaving you with propene.

$$CH_3CH_2CH_2Cl + c.NaOH \longrightarrow CH_3CH{=}CH_2$$

ISBN: 9780170260107

9 a i Acidified potassium permanganate

ii Oxidation

b i

Name: 2-chloropentane Structural formula: $CH_3CH(Cl)CH_2CH_2CH_3$	Name: 1-chloropentane Structural formula: $CH_2ClCH_2CH_2CH_2CH_3$

ii Addition

iii 2-chloropentane – Major product
1-chloropentane – Minor product

iv Markovnikov's rule states that the 'rich get richer', meaning the carbon in the double bond with the greatest number of hydrogens will preferentially gain the extra hydrogen over the other carbon in the double bond, so this forms the major product 2-chloropentane. The minor product is the reverse of this, with the carbon that is hydrogen rich gaining the chlorine atom. Much less of this product will form hence it is the minor product.

10 a **Classification:** Tertiary
Reason: The carbon attached to the chlorine atom is attached to three other carbons.
Classification: Secondary
Reason: The carbon attached to the chlorine atom is attached to two other carbons.

b The two molecules have the same molecular formula $C_5H_{11}Cl$ but a different arrangement in space, the requirement for them to be structural isomers.

c

```
     H   H   H   H   H             H   H   H   H   H
     |   |   |   |   |             |   |   |   |   |
Cl — C — C — C — C — C — H     H — C — C — C — C — C — H
     |   |   |   |   |             |   |   |   |   |
     H   H   H   H   H             H   Cl  H   H   H
```

1-chloropentane   2-chloropentane

```
    H   H   Cl  H   H                  CH3
    |   |   |   |   |                  |
H — C — C — C — C — C — H       CH3 — C — CH2 — Cl
    |   |   |   |   |                  |
    H   H   H   H   H                  CH3
```

3-chloropentane   1-chloro-2,2-dimethylpropane

$CH_2ClCH(CH_3)CH_2CH_3$ 1-chloro-2-methylbutane

$CH_3CH(CH_3)CH_2CH_2Cl$ 1-chloro-3-methylbutane

Chapter Three

Pre-test

Section 1: Rates of reaction

1 Catalyst, concentration, temperature and surface area.

2 Collision theory is the theory that particles must collide with one another in order for a chemical reaction to occur. If you increase the rate of a reaction you must then also be increasing the number of collisions that occur per second.

Section 2: Reversible reactions

2 Melting ice is an example of a reversible reaction because you can get all the ice back by refreezing it.

Section 3: Acids and bases

1 A substance with a pH less than 7.

2 A substance with a pH greater than 7.

3 This is called a neutralisation reaction as you end up with two neutral products: salt and water.

4 This is also a neutralisation reaction, however the products are carbon dioxide, water and a salt.

5 How acidic or basic a substance is.

6 An indicator is a substance that changes in colour depending on the pH of a solution. Examples of indicators include – universal, litmus, methyl blue and methyl orange, however there are many more.

3.1 – Rates of reaction

1 a Manganese dioxide is acting as a catalyst in this reaction. It is speeding up the rate of reaction without being used up in the process. A catalyst works by lowering the activation energy required for a reaction to occur by providing an alternative pathway for the reaction. It will increase the number of collisions occurring per second because there will be more reactants with enough energy to react at the beginning of the reaction.

b Marble chips have a smaller surface area than the powered calcium carbonate, therefore there will be fewer particles available for the reaction at the beginning. There will then be fewer collisions per second occurring and so a slower rate of reaction.

c This is because the 2 $molL^{-1}$ hydrochloric acid has a greater concentration of acid particles available to react with the calcium carbonate. This will increase the number of collisions occurring per second as there are more acid particles available at the beginning of the reaction, increasing the rate of the reaction.

d In the fridge the temperature is lower and so the particles will have less kinetic energy. This will decrease the number of successful collisions occurring per second and so slow the rate of reaction.

Experiment 1 - How do catalysts alter the rate of reaction?

RESULTS

Reaction	Observations	Gas test
Hydrogen peroxide without manganese dioxide	No visible reaction	
Hydrogen peroxide with manganese dioxide	Bubbles of oxygen form gas very rapidly.	Glowing splint is relit by the oxygen gas.

Reaction	Observations
Zinc and sulfuric acid without copper	Some bubbles occur around the zinc metal.
Zinc and sulfuric acid with copper	Vigorous bubbles occur about the zinc metal.

QUESTIONS

1 You could weigh it accurately beforehand, filter it out once the reaction has finished and then weigh it again and compare the masses.

2 It is a catalyst which speeds up a rate of a reaction by providing an alternative pathway for the reaction to occur upon.

3 $Zn + H_2SO_4 \xrightarrow{Cu} ZnSO_4 + H_2$

Investigation 1: How does concentration affect the rate of a reaction?

QUESTIONS

1 The time taken is increased.

2 $Mg + 2HCl \longrightarrow MgCl_2 + H_2$

3 At the beginning of any reaction there are more reactant particles available for collisions to occur with, so a faster rate of reaction as there will be more frequent collisions per second. Near the end the reaction will slow down as there are fewer reactant particles available for collisions to occur with, therefore less frequent collisions.

ISBN: 9780170260107

Investigation 2: How does the surface area affect the rate of a reaction?

QUESTIONS

1 The powdered form as it had a greater surface area and so there were more particles available for collisions to occur with, therefore more collisions per second and a faster rate of reaction.

3 Cutting up vegetables smaller before cooking them, makes them cook much faster.

Experiment 2: Temperature and rate of reaction

CONCLUSION

As the temperature is increased the rate of reaction ***increases***. This is because the particles are moving ***faster*** in solution, therefore there are more ***collisions*** occurring. This ***increases*** the rate of reaction.

QUESTIONS

1 The time taken decreases with the increase in temperature, as the rate of reaction increases. This is because the particles have more kinetic energy and so will collide more frequently, leading to more frequent successful collisions.

2 Keep the reaction in cooler conditions.

Checkpoint 1 – Rate of reaction

1 You could measure the amount of hydrogen gas released, by collecting the gas by downward displacement or collecting the gas in a balloon, or you could measure the time taken for the piece of magnesium to disappear.

2 **a** 1.5 grams.
b 25 minutes
c In the 5 minutes because the curve is the steepest at this time.

3 The particles must be in the correct orientation and hit each other with enough force in order to be successful.

3.2 – Reversible reactions

1 You could get the reactants back.

2 No, because part of the iron has completely reacted with the oxygen in the air to form the red iron oxide.

3 No, because you can't un-burn the toast and get back the bread.

4 Yes, because the ice can be re-frozen.

5 The sublimation of iodine from solid to gas.

3.3 - Equilibrium constants, K_c

3 $K_c = [OH^-][HSO_3^-]/[SO_3^{2-}]$

4 $K_c = [CO_2][H_2]/[CO][H_2O]$

5 $K_c = [Fe^{3+}]/[Fe^{2+}][Ag^+]$

6 $K_c = [CO_2]^2/[CO]^2[O_2]$

7 $K_c = [O_3]^2/[O_2]^3$

8 $K_c = [CuCl_4^{2-}]/[Cu(H_2O)_6^{2+}][Cl^-]^4$

9 $K_c = [CO]^2/[O_2]$

10 $K_c = [NO]^4[H_2O]^6/[NH_3]^4[O_2]^5$

3.4 – Le Chatelier's principle

1 **a** Increasing the temperature would shift the equilibrium towards the reactants. This is because the system would shift to try and minimise the change made to the system by moving in the endothermic direction of this reaction in order to remove the increase in heat. K_c will decrease as there will be more reactants to products in the new equilibrium.

b As there are 9 moles of gas on the reactant side of this reaction and 10 on the product side the equilibrium will shift towards the side with the fewest moles in order to minimise the change made to the system by the increase in pressure. As the side with the fewest moles of gas is the reactant side the equilibrium will shift in this direction. K_c will remain unchanged as changes in pressure do not alter its value.

c Adding a catalyst will increase both the forward and backward reaction rates equally and so will have no effect on either the K_c value or shift the equilibrium's position.

d Decreasing the ammonia, NH_3 concentration will force the equilibrium to shift towards the reactants in order to minimise the change made to the system. This will force more ammonia to be formed to replace the ammonia that was lost. No change will be made to the K_c value.

2 **a** Increasing pressure will shift this equilibrium towards the products as there are only two moles of gas compared to the three on the reactants side. The equilibrium will shift in this direction in order to minimise the change made to the system, in order to lower the pressure of the system back down. No change to K_c will be observed.

b Adding more sulfur dioxide, SO_2 will cause the equilibrium to shift towards the products in order to minimise the change made to the system. This shift will reduce the excess sulfur dioxide to the system and form more products in the process. There will be no change to the value of K_c.

c Adding a catalyst will increase both the forward and reverse reaction rates equally without altering either the position or value of the equilibrium constant, however equilibrium will be reached more quickly.

d Decreasing the temperature will shift the equilibrium towards the products of the exothermic direction of this equilibrium in order to minimise the change made to the system. This is so the exothermic reaction will return some of the heat lost by the change in surrounding temperature. As the amount of products produced will increase the value of K_c will also increase and so a new K_c will be established with a higher value.

3 **a** As more of a reactant has been added to the system the equilibrium will shift towards the products in order to minimise the change made to the system. This is so the system can remove some of the excess $Cu(H_2O)_6^{2+}$ added, the colour will become more yellow as more of the yellow $CuCl_4^{2-}$ will be formed.

b As more of a reactant has been added to the system the equilibrium will shift towards the products in order to minimise the change made to the system. This is so the system can remove some of the excess Cl^- added, the colour will become more yellow as more of the yellow $CuCl_4^{2-}$ will be formed.

4 **a** Increasing the pressure will shift the equilibrium towards the products as there is only one mole of gas on the products compared to the three on the reactant side of the equation. The equilibrium shifts in order to minimise the change made to the system by removing some of the excess pressure in the system by reducing the number of moles present. As the only product is methanol there will be an increase in its overall concentration.

b Increasing the concentration of carbon monoxide will shift the equilibrium towards the products in order to minimise the change made to the system. This way the equilibrium removes the excess carbon monoxide added to the system, meaning an increase in the methanol concentration.

c Adding a catalyst will only increase the rate of the reaction in both directions and will have no affect on methanol's concentration.

5 **a** When the temperature is increased the equilibrium will shift in the endothermic direction of this equilibrium which

ISBN: 9780170260107

in this case is towards the products. The system does this in order to minimise the change made to the system by removing some of the added heat. As the purple colour comes from the iodine one of the reactants there will be less purple colour in the new equilibrium.

b When the temperature is decreased there will be less hydrogen iodide formed as the equilibrium will shift towards the reactants the exothermic direction of this reaction in order to minimise the change made to the system by adding more heat that was lost to the surroundings of the system.

Experiment 3: Observing changes to a chemical equilibrium

1 Colourless.
2 Pale yellow.
3 Blood red
4 $Fe^{3+}_{(aq)} + SCN^-_{(aq)} \rightleftharpoons FeSCN^{2+}_{(aq)}$
5 When more KSCN was added the system shifted towards the products in order to minimise the change made to the system by removing the excess SCN^- ions. The colour went darker red to reflect this.
6 The system shifted towards the reactants in order to minimise the change made to the system by replacing some of the lost Fe^{3+}. The solution went paler red to reflect this.

Investigation 3: Explaining the changes in concentration made to the chromate/dichromate equilibrium

1 It went orange.
2 When you added more reactants the system moved in order to correct this change and made more dichromate, making it more orange in colour.
3 No it would not as only a change in temperature alters the ratio of products to reactants and so the K_c value remains unchanged.

Investigation 4: Explaining the changes in temperature made to the cobalt ion equilibrium

1 The equilibrium shifts in the endothermic direction of this equilibrium making more blue $[Co(H_2O)_6]^{2+}$ in order to minimise the change made to the system.
2 The system shifts towards the reactants in order to reduce the excess heat added to the system.
3 K_c will become smaller as there will be a greater ratio of reactants to products formed in the new equilibrium that is established.
4 The enthalpy value in the forward direction is therefore negative and so exothermic since when the system is heated it shifts in the endothermic direction or towards the reactants. We know this due to the change in colour observed.

3.5 – NCEA-style Le Chatelier's principle questions

1 **a** When the temperature of the system is decreased the equilibrium system will shift in the exothermic direction in order to minimise the change made to the system. As this is the forward direction more sulfur trioxide will be formed. This happens so heat that is lost to the system is regained by more of the exothermic reaction occurring. As the concentration of the products will increase, K_c will also increase in value.

b When the concentration of sulfur dioxide is halved the equilibrium will shift towards the reactants in order to minimise the change made to the system by replacing the lost sulfur dioxide. K_c will remain unchanged as concentrations of reactants do not alter the value of it.

2 **a** The equilibrium will shift towards the side with the fewest moles of gas in order to minimise the change made to the system. In this case the reactant side has only one mole of gas, whereas the products side has two so the equilibrium will shift towards the reactants in order to lower the pressure in the system by reducing the number of moles present. K_c will not change as pressure changes do not alter the value of K_c.

b A catalyst provides an alternative pathway for a reaction to occur; it increases both the forward and reverse reaction rates equally so the equilibrium will be established earlier but there will be no shift in its position or in the value of K_c.

3 **a** When the concentration of copper ions is doubled the equilibrium will shift in order to minimise the change made to the system, in this case towards the reactants in order to remove the excess copper ions added to the system. K_c will not be altered as concentration changes will not alter its overall value.

b When the temperature of the system is increased the equilibrium will shift in order to minimise the change made to the system by shifting in the endothermic direction, in this case towards the reactants. The system does this as the endothermic reaction will remove some of that added heat. The new equilibrium established will have a smaller K_c value as more reactants will be formed to products.

3.6 – K_c calculations

1 $K_c = [N_2O_4]/[NO_2]^2 = [0.0201]/[0.102]^2 = 1.93$
2 $K_c = [SO_3]^2/[SO_2]^2[O_2] = [0.931]^2/[0.240]^2[1.28] = 11.8$
3 $K_c = [HI]^2/[H_2][I_2] = [0.175]^2/[0.0275][0.0290] = 38.4$
4 $K_c = [S_2][H_2]^2/[H_2S]^2 \quad 2.25 \times 10^{-4} = [2.20 \times 10^{-3}][H_2]^2/[5.01 \times 10^{-3}]$
$[H_2] = \sqrt{2.25 \times 10^{-4} \times [5.01 \times 10^{-3}]/[2.20 \times 10^{-3}]} = 0.140 \text{ mol L}^{-1}$

Checkpoint 2 – Equilibrium systems

1 **a** $K_c = [HI]^2/[H_2][I_2]$

b $64 = [HI]^2/0.100 \times 0.100$
$[HI] = \sqrt{64 \times 0.100 \times 0.100} = 0.800 \text{ mol L}^{-1}$

c With a high K_c value there will be more of the products HI than the reactants as a high K_c corresponds to a product favoured equilibrium.

d If the temperature of the system was doubled the equilibrium system would shift in order to minimise the change made to the system, it would shift towards the products in this case as that is the endothermic direction. It shifts in this direction to remove some of the extra heat added to the system. K_c will become larger because of this as the reaction will become even more product favoured.

e Doubling the pressure will cause the equilibrium to shift in order to minimise the change made to the system. In order to lower the pressure back down, the equilibrium will shift to the side with the fewest moles of gas. In this case both the products and reactants have two moles of gas, so there will be no shift. However equilibrium will be reached faster as the rate of reaction will increase with the increase in pressure. There will be no change to the K_c value.

f If we halve the concentration of hydrogen gas present the equilibrium will shift towards the reactants in order to minimise the change made to the system. In this way more hydrogen will be formed to replace what was lost. K_c will not change with a change in concentration.

3.7 – Writing acid-base reactions

1 and

3 **a** $HCl + H_2O \longrightarrow Cl^- + H_3O^+$ $\quad$ HCl/Cl^- and H_3O^+/H_2O both conjugate acid/conjugate base

ISBN: 9780170260107

b $HNO_3 + H_2O \rightarrow NO_3^- + H_3O^+$ HNO_3/NO_3^- and H_3O^+/H_2O

c $H_2SO_4 + H_2O \rightarrow HSO_4^- + H_3O^+$ H_2SO_4/HSO_4^- and H_3O^+/H_2O

d $CH_3CH_2COOH + H_2O \rightleftharpoons CH_3CH_2COO^- + H_3O^+$ $CH_3CH_2COOH/CH_3CH_2COO^-$ and H_3O^+/H_2O

e $HF + H_2O \rightleftharpoons F^- + H_3O^+$ HF/F^- and H_3O^+/H_2O

f $CH_3COOH + H_2O \rightleftharpoons CH_3COO^- + H_3O^+$ CH_3COOH/CH_3COO^- and H_3O^+/H_2O

2 and

3 a $NaOH_{(aq)} \rightarrow Na^+ + OH^-$ $Na^+/NaOH$

b $Mg(OH)_{2\ (aq)} \rightarrow Mg^{2+} + 2OH^-$ $Mg^{2+}/Mg(OH)_2$

c $NH_3 + H_2O \rightleftharpoons NH_4^+ + OH^-$ NH_4^+/NH_3 and H_2O/OH^-

d $CH_3COO^- + H_2O \rightleftharpoons CH_3COOH + OH^-$ CH_3COOH/CH_3COO^- and H_2O/OH^-

4 a $3HCl + Al(OH)_3 \rightarrow AlCl_3 + 3H_2O$

b $6HCl + Al_2(CO_3)_3 \rightarrow 2AlCl_3 + 3CO_2 + 3H_2O$

c $2HNO_3 + Zn \rightarrow Zn(NO_3)_2 + H_2$

d $2CH_3COOH + PbCO_3 \rightarrow Pb(CH_3COO)_2 + CO_2 + H_2O$

e $2CH_3COOH + Mg \rightarrow Mg(CH_3COO)_2 + H_2$

Experiment 4: The strengths of acids and their reactions

1 Colours in Universal indicator

Acid	Colour in Universal indicator	pH
Hydrochloric acid	*Red*	*1*
Ethanoic acid	*Orange*	*3*

2 Reaction with magnesium

Acid	Observations with magnesium
Hydrochloric acid	*Lots of effervescence, loud pop*
Ethanoic acid	*Some effervescence, soft pop*

3 Reaction with calcium carbonate (marble)

Acid	Observation with calcium carbonate
Hydrochloric acid	*Lots of effervescence, limewater goes milky rapidly*
Ethanoic acid	*Some effervescence, limewater goes milky slowly*

QUESTIONS

1 Hydrochloric acid is the strong acid and ethanoic acid is the weak acid, as the hydrochloric acid turned red in universal indicator and reacted more vigorously with the magnesium and calcium carbonate.

2 a $HCl + H_2O \rightarrow Cl^- + H_3O^+$

b $CH_3COOH + H_2O \rightleftharpoons CH_3COO^- + H_3O^+$

c $2HCl + Mg \rightarrow MgCl_2 + H_2$

d $2\ CH_3COOH + Mg \rightarrow Mg(CH_3COO)_2 + H_2$

e $2HCl + CaCO_3 \rightarrow CaCl_2 + CO_2 + H_2O$

f $2CH_3COOH + CaCO_3 \rightarrow Ca(CH_3COO)_2 + CO_2 + H_2O$

3 You would observe a similar result with the concentrated acid reacting more vigorously than the dilute acid as there are more acid particles present in the solution of the concentrated one, hence a faster rate of reaction.

4 The concentrated ethanoic acid would still be less vigorous than the hydrochloric acid due to ethanoic acid only partially dissociating in water, however you would still see a slower rate of reaction with the more dilute sample.

3.8 – Acid calculations

1 a pH = - log(1) = 0

b pH = - log(0.01) = 2

c pH = - log(0.0122) = 1.91

d pH = - log(1.45) = -0.161

e pH = - log(2.13) = -0.328

f pH = - log(2 x 0.478) = 0.0195

2 a $[H_3O^+] = 10^{-1} = 0.100$ molL^{-1}

b $[H_3O^+] = 10^{-4.55} = 2.82 \times 10^{-5}$ mol L^{-1}

c $[H_3O^+] = 10^{-3.26} = 5.50 \times 10^{-4}$ mol L^{-1}

d $[H_3O^+] = 10^{-6.66} = 2.19 \times 10^{-7}$ mol L^{-1}

e $[H_3O^+] = 10^{-0.554} = 0.279$ mol L^{-1} $[H_2SO_4] = 0.140$ molL^{-1}

f $[H_3O^+] = 10^{-1.27} = 5.37 \times 10^{-2}$ mol L^{-1} $[H_2SO_4] = 2.69 \times 10^{-2}$ mol L^{-1}

3.9 – Base calculations

1 a $[H_3O^+] = 1 \times 10^{-14}/0.1 = 1 \times 10^{-13}$ mol L^{-1} pH = $-\log[1 \times 10^{-13}] = 13$

b $[H_3O^+] = 1 \times 10^{-14}/0.0160 = 6.25 \times 10^{-13}$ mol L^{-1} pH = $-\log[1 \times 10^{-1}] = 12.2$

c $[H_3O^+] = 1 \times 10^{-14}/1.20 = 8.33 \times 10^{-15}$ mol L^{-1} pH = $-\log[8.33 \times 10^{-15}] = 14.1$

d $[H_3O^+] = 1 \times 10^{-14}/0.0126 \times 2 = 3.96 \times 10^{-13}$ mol L^{-1} pH = $-\log[3.96 \times 10^{-13}] = 12.4$

e $[H_3O^+] = 1 \times 10^{-14}/3 \times 2.00 = 1.66 \times 10^{-15}$ mol L^{-1} pH = $-\log[1.66 \times 10^{-15}] = 14.8$

2 a $[H_3O^+] = 10^{-8.66} = 2.19 \times 10^{-9}$ mol L^{-1} $[OH^-] = 4.57 \times 10^{-6}$ molL^{-1}

b $[H_3O^+] = 10^{-10.1} = 7.94 \times 10^{-11}$ mol L^{-1} $[OH^-] = 1.26 \times 10^{-4}$ molL^{-1}

c $[H_3O^+] = 10^{-14} = 1 \times 10^{-14}$ mol L^{-1} $[OH^-] = 1.00$ molL^{-1}

d $[H_3O^+] = 10^{-11.6} = 2.51 \times 10^{-12}$ mol L^{-1} $[OH^-] = 4.00 \times 10^{-3}$ molL^{-1} $[Cu(OH)_2] = 2.00 \times 10^{-3}$ mol L^{-1}

e $[H_3O^+] = 10^{-13.8} = 1.58 \times 10^{-14}$ mol L^{-1} $[OH^-] = 0.63$ mol L^{-1} $[Zn(OH)_2] = 0.315$ molL^{-1}

3.10 – Acid base explanations

1 The nitric acid will react more vigorously with the magnesium metal as it is a strong acid and so will completely dissociate in water, meaning all its acid particles will be available to react at the beginning of the reaction. Ethanoic acid is a weak acid so it will only partially dissociate in water to produce hydronium ions, therefore it will have a slower rate of reaction. As the equilibrium between the ethanoic acid and its hydronium ions shifts towards the products all the hydronium ions are reacted. Both reactions will still go towards completion though.

$$HNO_3 + H_2O \rightarrow H_3O^+ + NO_3^-$$
$$2HNO_3 + Mg \rightarrow Mg(NO_3)_2 + H_2$$
$$CH_3COOH + H_2O \rightleftharpoons H_3O+ + CH_3COO^-$$
$$2CH_3COOH + Mg \rightarrow Mg(CH_3COO)_2 + H_2$$

2 The sulfuric acid is a strong acid and so will completely dissociate in water to produce hydronium and hydrogen sulfate ions, therefore there will be plenty of ions available in solution to conduct a current. Since in order to conduct a current there must be free moving charged particles, in this case there will be ions.

$$H_2SO_4 + H_2O \rightarrow HSO_4^- + H_3O^+$$

Propanoic acid is a weak acid and so will only partially dissociate in water to produce the propanoate ion and the hydronium ion, therefore there will be less ions available to conduct a current and so it won't conduct as well as the sulfuric acid.

$$CH_3CH_2COOH + H_2O \rightleftharpoons CH_3CH_2COO^- + H_3O^+$$

 ISBN: 9780170260107

Checkpoint 3 – Acids and bases

1 Concentration refers to the amount of moles of acid or base added to a solution, whereas strength refers to how much an acid or base dissociates in water.

2

	Acid/Base	pH	Weak/Strong
A	Hydrochloric acid	1	Strong
B	Propanoic acid	3	Weak
C	Milk	6	Weak
D	Blood	8	Weak
E	Lime water/Calcium hydroxide	10	Strong
F	Sodium hydroxide	14	Strong

$$HCl + H_2O \longrightarrow Cl^- + H_3O^+$$
$$CH_3CH_2COOH + H_2O \rightleftharpoons CH_3CH_2COO^- + H_3O^+$$
$$Ca(OH)_{2\ (aq)} \longrightarrow Ca^{2+} + 2OH^-$$
$$NaOH_{(aq)} \longrightarrow Na^+ + OH^-$$

3 a $CH_3COOH + NaOH \longrightarrow NaCH_3COO + H_2O$

b $H_2SO_4 + 2KOH \longrightarrow K_2SO_4 + 2H_2O$

4 Window cleaners and oven cleaners are generally basic solutions. Soap is a base. Vinegar is used as a preserver and is an acid.

5 HCl

H_3O^+ Cl^-
Cl^- H_3O^+
H_2O H_2O
H_3O^+ OH^-
H_2O
Cl^- H_3O^+
H_3O^+ Cl^- H_2O

NH_3

NH_3 H_3O^+
OH^- NH_4^+
NH_4^+ OH^-
NH_4^+ OH^-
NH_3 H_2O

Exam-type questions

1 The manufacturers could increase the temperature as this will increase the kinetic energy of the particles, making them collide more frequently and with a greater force making more of the collisions successful. This will then therefore increase the rate of reaction.

The manufacturers could add a catalyst as this would lower the amount of activation energy required for the reaction to occur by providing an alternative pathway for the reaction to occur upon. This will mean more reactants will have enough energy to react and so a faster rate of reaction.

The manufacturers could increase the concentration of one of the reactants as then there will be more available particles for collisions to occur with and so more successful collisions will occur per second which will increase the rate of the reaction.

2 The magician could lower the concentration of one of the reactants which would mean there were less particles available to react with in the beginning, lowering the number of successful collisions and so therefore also lowering the rate of reaction.

The magician could lower the temperature surrounding the reaction, for example put the reaction vessel in ice. This will lower the kinetic energy of the reactants leading to less successful collisions occurring per second and so a lower rate of reaction.

3 a The variable being tested for in this investigation is concentration of the sulfuric acid. The greater the concentration of the acid used the more particles that are available in the solution to react with and so the faster the rate of reaction as more successful collisions will be occurring per second.

b You could increase the temperature as this will increase the kinetic energy of the particles making them collide more frequently and with a greater force making more of them successful. This will then therefore increase the rate of reaction.

You could add a catalyst as this would lower the amount of activation energy required for the reaction to occur by providing an alternative pathway for the reaction to occur upon. This will mean more reactants will have enough energy to react and so a faster rate of reaction.

Finally you could increase the surface area of the zinc metal, which would provide more particles available for a successful collision to occur with and so a faster rate of reaction as more successful collisions would occur per second.

4 a $K_c = [CO_2]$

b If you heat the equilibrium the system will shift to minimise the change made to it, by shifting towards the products as it will shift in the endothermic direction. It shifts in this direction as the endothermic reaction will take heat in from the surroundings therefore lowering the temperature back down. K_c will increase in value as there will be more products formed than before.

c A catalyst would speed up the rate of reaction in both directions, so the equilibrium will be reached faster whilst having no affect on the position value of the equilibrium. Therefore it will not alter K_c.

5 a $K_c = [N_2O_4]/[NO_2]^2$

b $K_c = [0.1]/[0.1]^2 = 10$

c It is product favoured since the value of K_c is greater than one.

d The equilibrium would shift towards the reactants to replace the lost nitrogen dioxide gas making the new equilibrium more brown in colour as more nitrogen dioxide will be formed. This is because the equilibrium will shift to minimise the change made to the system.

e The equilibrium would shift in order to minimise the change made to the system by the lowering of the temperature and so become less brown in colour. It will become less brown as the equilibrium will shift towards the products in the exothermic direction of this reaction in order to increase the heat that was lost. The brown colour is from the nitrogen dioxide which it is moving away from. K_c will become larger in value as the reaction will become more products favoured.

6 a $K_c = [FeCl_4^-]/[Fe^{3+}][Cl^-]^4$

b $8.00 \times 10^{-2} = [FeCl_4^-]/[0.220][0.801]^4$

$[FeCl_4^-] = 8.00 \times 10^{-2} \times [0.220][0.801]^4$

$[FeCl_4^-] = 0.00725\ molL^{-1}$

c The equilibrium would shift to correct the change made to the system, in this case that would be towards the reactants in order to replace the chloride that was lost. There would be no change to the value of K_c as concentration changes do not affect K_c.

7 a $K_c = [I_3^-]/[I^-][I_2]$

b I^-. This is because if K_c is a small number it must mean that there are more reactants to products, since reactants are on the bottom half of the K_c expression.

c $7.11 \times 10^{-2} = [0.156]/[[0.125][I_2]$ $[I_2] = 0.156/7.11 \times 10^{-2} \times 0.125 = 0.274\ mol\ L^{-1}$

8 Acid X is a strong acid which will dissociate completely in water to produce its conjugate base and hydronium ions, therefore it will have a fast rate of reaction with calcium carbonate as all the acid particles or hydronium ions will be free to react at the beginning of the reaction. We know acid X is a strong acid given that it has the same concentration as acid Y but a much lower pH.

$HX + H_2O \longrightarrow X^- + H_3O^+$
$HX + CaCO_3 \longrightarrow CaX_2 + H_2O + CO_2$

Acid Y is a weak acid as it has a much higher pH even though

ISBN: 9780170260107

it has the same concentration as acid X. This means it does not completely dissociate in water to produce its conjugate base and hydronium ions and so it will react much more slowly with calcium carbonate. It will still go to completion though because, as the hydronium ions are reacted with the calcium carbonate they are removed from the equilibrium with water forcing the equilibrium to go to completion.

$HY + H_2O \rightleftharpoons Y^- + H_3O^+$

$HY + CaCO_3 \rightarrow CaY_2 + CO_2 + H_2O$

9 **a** HCl conjugate acid with Cl^- the conjugate base; H_2O conjugate base with H_3O^+ conjugate acid.

b NaOH conjugate base with H_2O the conjugate acid; HCl is the conjugate acid with Cl^- the conjugate base.

c CH_3COOH is the conjugate acid with CH_3COO^- the conjugate base; H_2O is the conjugate base with H_3O^+ which is the conjugate acid.

d This is because ethanoic acid is a weak acid and so does not completely dissociate in water so is in an equilibrium. Whereas hydrochloric acid is a strong acid which does completely dissociate in water so its reaction goes to completion.

e As hydrochloric acid is a strong acid all its acid particles are available to react at the beginning of the reaction with the sodium hydroxide and therefore there will be a faster rate of reaction. Since ethanoic acid is in an equilibrium with water, not all of its acid particles are available to react with the sodium hydroxide, however as the hydronium ions are removed from the equilibrium the equilibrium shifts towards the products eventually going to completion. This will make the reaction with sodium hydroxide a lot slower but it will still go to completion.

10 Hydrochloric acid is a strong acid which will dissociate completely in water to produce its conjugate base and hydronium ions in the reaction shown below.

$HCl + H_2O \rightarrow Cl^- + H_3O^+$

This makes it a strong acid because it completely dissociates, which also makes it an excellent conductor as all of it is converted into ions which can conduct a current. (In order for an electric current to conduct there must be free moving electrons or ions like in this case to conduct a current.)

Ethanoic acid on the other hand is a weak acid – we can tell this from its pH value which is higher, despite having the same concentration as the hydrochloric acid. As it is a weak acid it does not completely dissociate in water to form its conjugate base and hydronium ions, hence the higher pH value. This means it will also be not as good at conducting as it has a lower concentration of free moving ions to conduct a current.

$CH_3COOH + H_2O \rightleftharpoons CH_3COO^- + H_3O^+$

ISBN: 9780170260107